/////////////

High-Speed
Analog-to-Digital Conversion

High-Speed
Analog-to-Digital Conversion

Michael J. Demler

Academic Press, Inc.
Harcourt Brace Jovanovich, Publishers
San Diego New York Boston London Sydney Tokyo Toronto

Portions of this text have been reproduced/adapted from IEEE Std 746-1984, *IEEE Standard for Performance Measurements of A/D and D/A Converters for PCM Television Video Circuits,* copyright © 1984, and IEEE Std 1057, *IEEE Trial-Use Standard for Digitizing Waveform Recorders,* copyright © 1989 by the Institute of Electrical and Electronics Engineers, Inc., with the permission of the IEEE Standards Department. In the event of any discrepancy between this version and the original IEEE Standard, the IEEE version takes precedence.

Academic Press, Inc.
San Diego, California 92101

United Kingdom Edition published by
Academic Press Limited
24–28 Oval Road, London NW1 7DX

Library of Congress Cataloging-in-Publication Data

Demler, Michael J.
    High-speed analog-to-digital conversion / Michael J. Demler.
        p.   cm.
    Includes bibliographical references and index.
    ISBN 0-12-209048-9
    1. Analog-to-digital converters.   I. Title.
    TK78887.6.D46   1991
    621.381'5322--dc20   90-49744
                                                                CIP

PRINTED IN THE UNITED STATES OF AMERICA
91 92 93 94   9 8 7 6 5 4 3 2 1

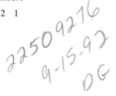

///////////////

This book is dedicated in loving memory of
my son Michael and to Mom, Dad, and Rosie
for all the years of support and encouragement.

Contents

6 / Test Methods for High-Speed A/D Converters

Preface

This book focuses on techniques for the high-speed conversion of analog signals into digital code. Devices and circuits that perform this task of data conversion are generally referred to as A/Ds (analog-to-digital converters), while the complementary function of digital-to-analog conversion is performed by D/As. The primary goal of this book is to provide a reference for practicing engineers who are working with data converters and to assist them in the selection and application of high-speed A/Ds. The subject matter grew out of the author's own experience in the design of high-speed A/D integrated circuits and his extensive contact with the actual users of these devices.

Paralleling the more widely heralded advances in purely digital devices such as microprocessors and memory chips, A/Ds have quietly followed a similar evolution. From the early circuit boards assembled with discrete components, today's integrated circuit technology enables the fabrication of complete single-chip data acquisition systems containing tens of thousands of transistors and precision networks of resistors and capacitors. The complementary development of digital signal processor (DSP) technology has increased the use of A/Ds as the interface for the acquisition and measurement of the many signals that exist in the real world as spectrums of analog waveforms.

The state of the art in A/Ds includes many different types of devices, but they can generally be grouped into three categories based on their architecture and performance: *high-resolution; high-speed,* or *flash;* and *subranging,* or *two-pass.*

High-resolution A/Ds are generally built using serial architectures that most often determine one bit at a time in a sequence of measurements. Examples of such devices include monolithic 16-bit devices and up to 22-bit hybrid components.

High-speed, or flash, A/D integrated circuits employ parallel techniques from which all the bits are determined simultaneously in a single measurement. Devices are now available that provide 8 bits of resolution at sampling rates as high as 500 megasamples per second and 10 bits of resolution at rates greater than 50 megasamples per second.

Subranging, or two-pass, A/Ds combine both serial and parallel techniques in order to extend high-speed performance to higher resolution than could be obtained with an entirely parallel approach. Examples of this architecture exist in hybrid devices providing up to 12 bits of resolution at 10 megasamples per second.

Although the label "high-speed" could be applied to an A/D in any of these categories as an indication of relative performance, it is not the intention of this book to engage in such competitive comparisons. The core of this book is based on the author's belief that the real breakthroughs in A/D speed are largely due to the rapid advances that have been made in flash converter technology. Unlike all other types of data converters, flash A/Ds are only practical in monolithic integrated circuit form. For designers of high-speed A/D systems, this book will provide vital information on the details of the parallel architectures that form flash A/Ds and also describe the subranging configurations that rely on flash A/Ds.

Although the main objective of this book is to aid in the application of A/D components and architectures where the highest possible conversion speed is required, design engineers need more than a circuit "cookbook" to build high-speed data converter systems. It is important to understand the internal structure of these devices because fundamental differences in technology and design eventually determine the proper choice for each application. Coverage of some of the details of the IC design and fundamental circuit theory is included in order to point out the salient features of the various types of A/Ds that are available. References are provided for those interested in exploring the fine points in more detail.

This book makes selection and application of A/Ds easier by breaking through the barriers created by specmanship and by conveying a clearer understanding of manufacturers' data sheets and the devices they represent. Every specification is intended to model the error sources that exist in real A/Ds. Rather than simply providing a list of definitions for the various parameters that are used, the mathematical derivation of these models is supplied. Without resorting to advanced mathematics purely to expound on theory, these derivations are intended to illustrate the underlying concepts in a manner that any engineer can follow. This background information is vital in order for designers to form direct comparisons where no single standard exists among the various manufacturers.

Chapter 1 sets a foundation for understanding A/Ds by reviewing the architectures that are most frequently utilized in the different categories of performance. This survey will assist the reader in identifying the type of A/Ds to consider for various applications.

Chapter 2 focuses on high-speed A/Ds by exploring the details of the functional blocks that make up flash A/Ds. The flash, or fully parallel, architecture provides the fastest possible technique for A/D conversion. Differences between the CMOS and bipolar implementations of flash A/Ds are emphasized.

Chapter 3 explains the purpose and importance of the various high-speed A/D specifications. A technical definition and derivation of each parameter will help the reader develop a clearer understanding of manufacturers' data sheets.

Chapter 4 describes the support circuits that are required to fully exploit the performance of flash A/Ds. Selection criteria are provided for ancillary ICs such as op-amps, buffers, and voltage

references. Tips are given for improving the performance of flash
A/Ds through proper application of peripheral components.

Chapter 5 gives some examples of application circuits that
utilize flash A/Ds and contains ideas on exploiting the flexibility
of these devices as building blocks. Subranging techniques that
extend high-speed conversion to higher resolution are also introduced.

Chapter 6 describes the testing methods that can be used to
evaluate high-speed A/D performance. Techniques for engineering characterization as well as manufacturing test are discussed.

The bibliography provides an extensive list of references from
the author's files that will allow the reader to explore some of the
A/D conversion topics in more detail. This material from many
different sources has contributed greatly to the author's expertise
and includes the key contributions that have been made to the
state of the art in A/D conversion.

Michael J. Demler

1

A/D Converter Architectures

There is no universal A/D that is best suited to all situations, but there should be an optimum approach to address the needs of each application. Misapplication of an A/D can occur when the objectives of the user of a certain device are inconsistent with those of its designer. To alleviate such problems, this chapter will provide an overview of the most popular architectures that are utilized to implement A/D converters. In block diagram format, functional descriptions will explain the key features of each type of A/D. By understanding the fundamental differences in the A/D types and the various advantages and disadvantages of each approach, the selection process can be refined to consider only those A/Ds that are likely to provide a good fit for the intended application.

Type I: Serial A/Ds

The common feature of all serial A/Ds is that they embody an algorithm which attempts to iteratively minimize the difference between the input signal and an analog approximation that is created proportional to some accurate reference source. In each comparison, or conversion cycle, at most 1 bit is determined and accumulated to form the final N-bit result. Said another way, the signal passes at least N times through a set of the A/D's basic functional blocks in all serial converters. This does not necessarily imply that serial approaches cannot be used to provide high speed, however, as will be seen from the descriptions that follow. The serial-type A/Ds include the widest variety of performance of any architecture, since this category encompasses the slowest types and those with the lowest power as well as the highest resolution of any A/D architecture.

Ramp or Integrating A/Ds

The basic architecture of an integrating A/D is shown in Fig. 1-1. This example represents the dual-slope technique, which is predominantly used in low-speed applications such as panel meters and digital volt meters. In such cases the high accuracy that can be achieved with this approach is of primary importance.

For the dual-slope A/D, quantization of the input signal proceeds in three phases. In the first phase a feedback loop is closed around the analog signal path while the input is set to zero. This cycle performs an autozero function, since any residual

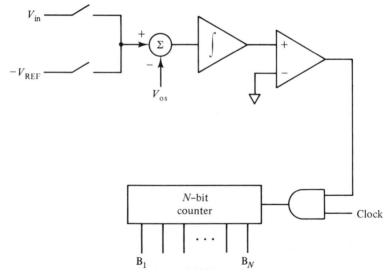

Figure 1-1   Architecture of dual-slope A/D converter.

errors in the signal path, such as offset in the integrator, are stored on a summing junction where they can be subtracted from the input.

In the second phase of conversion the analog signal is switched to the input of an integrator while the counter, which is cleared in the autozero phase, is allowed to count up to full scale. When the counter overflows, the integration of the input signal is stopped. Since a fixed number of clock cycles is always used (2^N for the N-bit counter), the charge accumulated on the integrating capacitor will be directly proportional to the amplitude of the input signal.

In the final phase of conversion an accurate reference voltage of opposite polarity to the analog signal is applied to the input of the integrator. The integration of this DC reference voltage will cause the integrator output to decrease linearly, as in the ramp shown in Fig. 1-2. During this interval the counter is again allowed to run up from zero. When the comparator input reaches the autozeroed level, where the conversion began, its output switches off the clock to the counter and terminates the conversion.

To illustrate how the quantized value of the input signal is obtained, the following equations describe the state of the integrator at the end of the conversion.

$$\int_0^{2^N \cdot T_{CLK}} v_{in} \, dt - \int_0^{T_c} V_{REF} \, dt = 0$$

$$\int_0^{2^N \cdot T_{CLK}} v_{in} \, dt = V_{REF} \cdot T_c$$

$$= V_{REF} \cdot COUNT \cdot T_{CLK}$$

For a DC input:

$$\frac{v_{in}}{V_{REF}} = \frac{COUNT}{2^N}$$

The high degree of accuracy in the quantization process results because, at least in this ideal case, the digital output is directly proportional to the integrated input signal. The final state of the

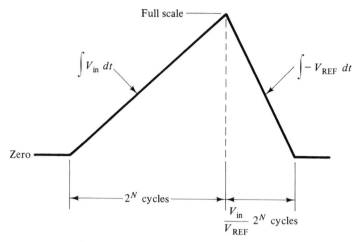

Figure 1-2   Dual-ramp conversion cycle.

counter depends only on the reference voltage and the degree to which the interval between clock pulses is constant. Very high stability reference circuits can be chosen, and it can also be assumed that clock instabilities will be insignificant over the duration of the conversion process. Accuracy of other components is not critical, since the reference and the input are both applied to the same signal path and autozeroing can be employed.

Limitations in the speed of this architecture are obviously associated with the number of bits in the counter. To digitize a full-scale signal (neglecting autozeroing time), 2^{N+1} clock cycles are necessary. The number of clock cycles doubles with each additional bit of resolution. One step that can be taken to speed up the process is to use a triple-ramp approach. In this technique two different rates of reference integration are employed. One counter can be used at higher speed for the coarse conversion to determine the MSBs, crossing over to a slower speed counter for the fine resolution required in the LSBs.

Delta-Sigma A/Ds

The delta-sigma (or sigma-delta) A/D is increasing in popularity for audio-band signal processing and low-frequency measurement applications. This is due to its ability to be easily integrated

on a chip that is predominantly made up of complex digital circuitry, without requiring more expensive precision analog components. There can also be speed advantages compared to other
integrating architectures. A typical implementation is shown in
Fig. 1-3.

Delta-sigma A/Ds are based on the translation of high-speed,
low-resolution samples of a band-limited input signal into a
higher resolution, lower speed digital output. Because of the
high-speed sampling operation relative to the limited input frequency, this type of A/D is usually referred to as an "oversampling" architecture. It is often implemented in a CMOS (complementary metal oxide semiconductor) process which allows the
signal processing blocks to be constructed using switched-
capacitor sampled-data techniques. The input samples are
summed in an integrator that feeds its output to a comparator,
which can be considered a 1-bit A/D. The comparator makes
decisions at the high-speed clock rate as to whether the integrator
output is positive or negative. These decisions are then fed serially to a digital filter and also used to control a 1-bit D/A in the
feedback loop, which outputs either +REF or −REF.

At first glance the delta-sigma A/D appears very similar to the
previous integrating A/D. However, by applying feedback from

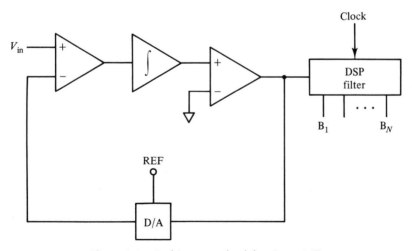

Figure 1-3   Architecture of a delta-sigma A/D.

the D/A and the differential amplifier, the loop forms a servo function, which tends to drive the integrator voltage toward zero. For large positive inputs, the initial integrator output will also be positive. A logic 1 is recorded at the comparator output, feeding back +REF to be subtracted from the input. A series of 1s will continue to result until the integrator output crosses the comparator threshold. As the cycle continues, the average value of the D/A output will become equal to the unknown input voltage. This value will also be represented by averaging the digital bit stream from the 1-bit A/D. The example in Fig. 1-4 illustrates this process.

For the example in Fig. 1-4, it is assumed initially that the integrator and D/A outputs are zero. A full-scale voltage of 1 V is used. The first sample stores 0.8 V at the integrator output. This

Example: V_{in} = 0.8 V

Sample	Differential amplifier	Integrator	Comparator	D/A
1	0.8	0.8	1	+1
2	−0.2	0.6	1	+1
3	−0.2	0.4	1	+1
4	−0.2	0.2	1	+1
5	−0.2	0	0	−1
6	1.8	1.8	1	+1
7	−0.2	1.6	1	+1
8	−0.2	1.4	1	+1
9	−0.2	1.2	1	+1
10	−0.2	1.0	1	+1
11	−0.2	0.8	1	+1
12	−0.2	0.6	1	+1
13	−0.2	0.4	1	+1
14	−0.2	0.2	1	+1
15	−0.2	0	0	−1

$$\text{Average D/A output} = \frac{9(+1) + 1(-1)}{10} = 0.8$$

Figure 1-4   Sampling sequence in a delta-sigma A/D.

positive voltage produces a logic 1 from the comparator and, subsequently, a 1-V output from the D/A. The next sample from the difference amplifier will produce -0.2 V, causing the integrator to ramp down to 0.6 V. The sequence of samples will produce a pattern of 1s and 0s from the comparator output, which in this case repeatedly returns the integrator output to zero every 10 clock cycles. As shown in the figure, by averaging the D/A output over this period, a voltage equal to the input is generated.

The speed advantage of this architecture is that the oversampling rate to achieve a given resolution is much less than the 2^{N+1} counts in the dual-ramp approach. As an example, one such device, the CSZ5316 from Crystal Semiconductor, puts out a new 16-bit word for every 128 samples. As will be discussed further in Chapter 3, all A/Ds produce noise in the quantization process that results from their having finite resolution. This noise is constant over the Nyquist-limited bandwidth of one-half the sampling rate. By utilizing a high degree of oversampling, this noise is effectively spread over a much wider bandwidth than would result from using the minimum sampling rate on the signal of interest. This process, which is described more fully in Chapter 5, increases the effective signal-to-noise ratio and hence the usable dynamic range. The digital filter at the output can attenuate the remaining noise that results from the quantization process.

The complexity of the required digital filter is one of the factors that limit the promise of monolithic delta-sigma A/Ds. In theory, the capability of implementing these A/Ds using a relatively simple digital process should result in lower cost. However, the large amount of digital circuitry often results in larger, more complex ICs that have lower yield and are more costly to produce. In order to realize the digital functions, an expensive VLSI process may be required. A digital process is also not capable of implementing the supporting analog functions required in measurement applications, such as S/Hs (sample/holds) and instrumentation amplifiers. Adding such components negates any cost savings which may be realized. Because of these factors, delta-sigma A/Ds will primarily be used in telecommunications and other digital audio applications.

Successive Approximation A/Ds

Without a doubt the most widely implemented A/D conversion algorithm is the serial method known as successive approximation. Devices that use this method exist with a wide range of conversion speed and eight to sixteen bits of resolution. The basic architecture is shown in Fig. 1-5.

The successive approximation algorithm is basically an intelligent tree search through all possible quantization levels, where each conversion step selects the next branch to follow based on the result of the previous estimate. This search path is shown in Fig. 1-6 for a 3-bit A/D. As a starting point, the first approximation is made exactly at the midscale point of the A/D, after which half of the possible solutions can be eliminated. In the implementation of Fig. 1-5, this is realized by introducing a "1" into the first bit of the shift register, which also sets the MSB of the successive approximation register (SAR) to the high state.

Since the MSB of the D/A has a weight of 2^{N-1}, or one-half full scale, in the first step a midscale signal is generated at the comparator input. A voltage level can be compared directly, but it is also common to sum current from the D/A output with current generated by the analog input. If the result of the MSB test is that

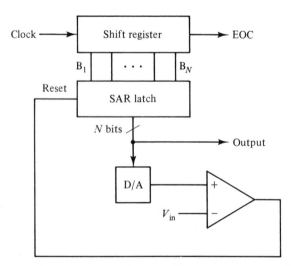

Figure 1-5   Architecture of a successive approximation A/D.

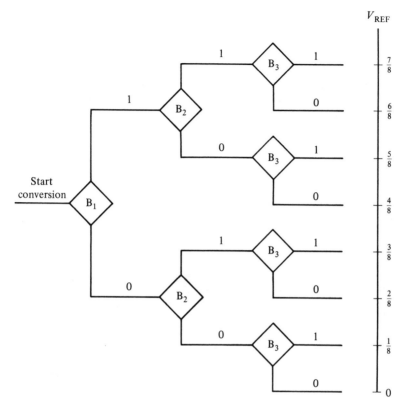

Figure 1-6   Search path for a 3-bit successive approximation A/D.

the D/A output is less than the input, then the first bit is left high
and the search will continue above midscale. If the opposite
condition occurs, the MSB is reset to "0" and the next test is
made at 1/4 of full scale. The procedure continues through the
lower bits, with the shift register enabling one bit at a time in the
SAR. A total of N comparisons are always made, and one bit of
the result is established in each approximation cycle.

The linearity of the successive approximation A/D is determined entirely by the performance of the D/A. Since N cycles are
needed to determine the result, a S/H circuit must be used to
keep the input signal stable to the desired accuracy while each bit
is evaluated. Variations equivalent to a fraction of an LSB would
cause a much larger error if comparisons in the MSB cycles were

corrupted. This would be equivalent to searching on the wrong branch of the tree—the right answer will never be found. The linearity of the S/H is also important to the overall system.

Conversion speed in the successive approximation A/D depends primarily on the settling time of the D/A and comparator. Logic delays also contribute to the total conversion time. To speed up the acquisition and processing of data, some successive approximation A/Ds provide a serial data output of each bit as it is determined. For data paths that are narrower than the A/D output, the MSBs and LSBs can be made available in separate bytes as they are determined.

There are other more sophisticated techniques for shortening the conversion time of successive approximation A/Ds, one of which is known as *progressive clocking*. This technique takes advantage of the fact that the quantization error gets progressively smaller as each bit in the successive approximation is determined. The D/A and comparator must settle to the same resolution for each bit, but the amount of voltage or current that the D/A output must slew at each step reduces by half. If the dynamic characteristics of the D/A are measured, the actual amount of settling time required at each step can be determined. A variable rate clock can then be used with a progressively shorter conversion cycle for each bit.

A simplified method of generating a progressive clock is shown in Fig. 1-7. In this example for an 8-bit A/D, the master clock must be divided by two to achieve the desired resolution in the MSBs. A total of 16 clock periods would be required for a complete conversion with a fixed clock rate. To adjust the conversion time of the LSBs, a counter is used as the sequencer of a simple state machine. The logic gates decode the state of the counter, which is cleared at the start of conversion. For the first four bits, the Q_A output is passed to the A/D clock, providing the divide-by-two function. When the counter reaches the eighth state, decoding of Q_D causes the A/D clock rate to double, passing the system clock to the output. Only 11.5 clock periods are now required to complete the conversion, an increase in speed of 28%. Variations on this clocking scheme are easily implemented by modifying the state decoding circuitry.

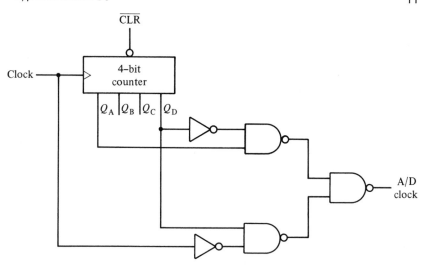

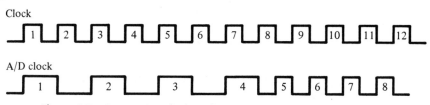

Figure 1-7   Progressive clocking for successive approximation A/Ds.

If finer resolution in the bit timing is desired, more sophisticated state machines can be used. A 1-bit-wide memory can be programmed to set the state of the A/D clock in increments of the master clock period. Larger counters are required to address the total number of states, with a higher ratio of system clock to A/D clock than in the simplified example. By stringing together a varying sequence of 1s and 0s, the cycle time of each bit in the A/D can be adjusted. Reductions in conversion time of as much as 40% have been reported with this technique. (For more details on this approach, see the references in the bibliography.)

Bit-Serial Pipelined A/Ds

The bit-serial pipelined architecture represents the upper limit on throughput rate from any A/D that quantizes one bit per conversion cycle. It is an intermediate approach between the successive approximation A/D and the subranging architecture that is described later. In the bit-serial pipeline there are N conversion stages, just as in the successive approximation A/D. However, once the first sample is completely quantized, there is a delay of only one cycle to acquire the next result. This is accomplished by resolving one bit at each stage and then passing the result on to the next stage for further processing. Once the result is passed on, each stage is free to process the next signal coming down the pipe. Parallel processing results in a throughput rate that is N-times faster than the equivalent successive approximation A/D.

Figure 1-8 is a simplified representation of such an architecture. Only three stages are shown, but the resolution can be extended by simply adding more of the identical processing blocks. Each block is capable of resolving one level of the same decision tree as the successive approximation A/D. The first stage, which follows an input S/H, compares the input signal to the midscale voltage represented by $V_{REF}/2$, which resolves the MSB. The subsequent stages contain a comparator, which is preceded by a S/H amplifier with a gain of 2. Switches select either the same input from the previous stage or the modified signal from the summing block. Not shown is the digital logic which assembles each bit into the complete digital output word.

Depending on the MSB result, the second stage must make decisions at either the 3/4 V_{REF} or 1/4 V_{REF} levels. The equations below describe how this is done. If MSB = 1, the following test is made:

$$2 \cdot (V_{in} - V_{REF}/2) \overset{?}{>} V_{REF}/2$$

which, by rearranging terms, is equivalent to

$$V_{in} \overset{?}{>} V_{REF}/2 + V_{REF}/4 = 3/4\ V_{REF}$$

When the MSB = 0, the input from the previous stage is fed forward to form the following test:

$$2 \cdot V_{in} \overset{?}{>} V_{REF}/2$$

After the first two levels of the successive approximation algorithm have been implemented, to complete the analysis, there

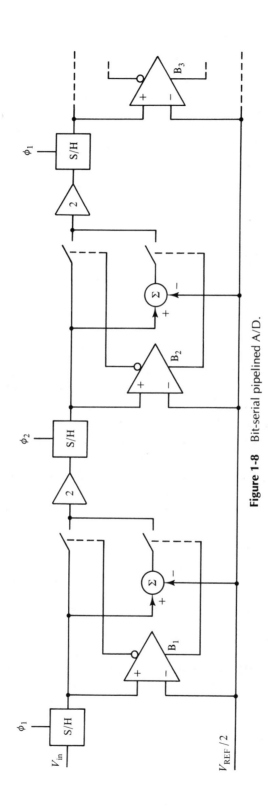

Figure 1-8 Bit-serial pipelined A/D.

are four decision levels which must be resolved at the third stage. If the second MSB=1, the possible comparisons are

$$\text{(I)} \quad 2 \cdot [2 \cdot (V_{in} - V_{REF}/2) - V_{REF}/2] > V_{REF}/2$$
$$\text{(II)} \quad 2 \cdot [2 \cdot (V_{in}) - V_{REF}/2] > V_{REF}/2$$

which are equivalent to

$$\text{(I)} \quad V_{in} > 7/8 \, V_{REF}$$
$$\text{(II)} \quad V_{in} > 3/8 \, V_{REF}$$

If the second MSB = 0, the decision levels are

$$\text{(III)} \quad 2 \cdot [2 \cdot (V_{in} - V_{REF}/2)] > V_{REF}/2$$
$$\text{(IV)} \quad 2 \cdot [2 \cdot (V_{in})] > V_{REF}/2$$

which simplify to

$$\text{(III)} \quad V_{in} > 5/8 \, V_{REF}$$
$$\text{(IV)} \quad V_{in} > 1/8 \, V_{REF}$$

The key feature of the bit-serial pipelined A/D is that each stage successively implements one level of the successive approximation algorithm. While the process for any single conversion still takes N cycles, the S/H and comparison circuitry actually allow N conversions to proceed simultaneously, resulting in one output for each cycle. The main limitation of this approach is the gain and offset errors that are introduced by processing the signal in each stage. Without some form of correction, these errors will propagate and multiply forward in the pipeline. To reduce the number of function blocks and, hence, the error sources, the concept can be extended by increasing the number of bits resolved in each stage. This will be explored together with digital error correction in the discussion of subranging A/Ds.

Algorithmic A/Ds

The elements of the processing blocks in the pipelined A/D can be rearranged slightly to implement the same conversion algorithm in a single stage. This architecture is referred to as an algorithmic or recirculating A/D. Rather than being passed through a pipeline, the signal is instead processed and fed back to test each bit serially. The benefits of such an approach are the reduction of components and the elimination of errors caused by the mismatch of multiple stages. Errors that occur can be compensated in a single stage and will be identical for each bit in the conversion sequence, resulting in improved linearity. This rearranged architecture is shown in Fig. 1-9.

The first step in the process is to store the signal in the S/H with a gain of 2. Comparison to the V_{REF} voltage resolves the MSB. In the subsequent steps, if the comparator output is "1," the reference voltage is subtracted from the comparator input and the result is fed back to the S/H. Otherwise, when the comparison is "0," the previous input signal is doubled and again compared to the reference voltage. By examining the process for determining the second MSB, it can be seen that the same decision points are implemented as in the feed-forward pipelined case mentioned earlier.

When MSB = 1:

$$\text{compare } 2 \cdot (2 \cdot V_{in} - V_{REF}) > V_{REF} \quad \text{or} \quad V_{in} > 3/4 \cdot V_{REF}$$

When MSB = 0:

$$\text{compare } 2 \cdot (2 \cdot V_{in}) > V_{REF} \quad \text{or} \quad V_{in} > 1/4 \cdot V_{REF}$$

The analog voltage will then be recirculated around the loop for each subsequent bit until the desired resolution is achieved.

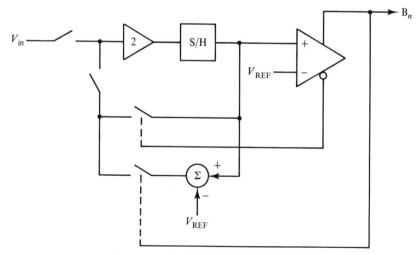

Figure 1-9 Algorithmic A/D converter.

Type II: Parallel A/Ds

The parallel A/D architecture, commonly referred to as flash A/D conversion, provides the fastest possible approach to quantizing an analog signal. One digitized sample is produced to N-bit resolution in each conversion cycle. This architecture relies on a "brute force" approach, where all of the possible quantization levels are simultaneously compared to the analog input signal. Figure 1-10 represents the architecture of a generic flash A/D.

To test all of the possible quantization levels of an N-bit A/D, 2^N-1 comparators are required. An additional comparator is often used to indicate the overflow condition at full scale. A reference decision level for each comparator is typically generated by voltage division of a precision reference. This requires 2^N+1 resistors to implement. After each comparator is strobed to sample the input, a bank of 2^N latches stores the results in a form that is referred to as a *thermometer code*.

In order for an analog signal to be quantized, its voltage level must be within the end points of the reference voltage divider. Comparators for which the analog input level is greater than the respective reference levels will output digital logic 1 after being strobed. Likewise, the comparators that have reference levels that are greater than the analog input will output logic 0 as a result of the comparison process. The appearance of the 2^N-bit digital word that results from the simultaneous comparisons gives rise to the thermometer analogy. A continuous string of 1s should appear up to the quantization level that is nearest to the input level. Beyond this point, a continuous string of zeros results. The height of the string of 1s can be read, just like mercury in a thermometer, to yield the quantized measurement of the analog input.

In actual implementation, digital logic is used to evaluate the results of adjacent comparators in order to find the 1/0 boundary condition. The thermometer decoder will produce only one output that is true, thus providing a unique digital code to the $2^N:N$ priority encoder in the second stage. The active signal acts as an address line to the second encoder, which is essentially a ROM, enabling the output of a single N-bit word equivalent to the value of the detected quantization level.

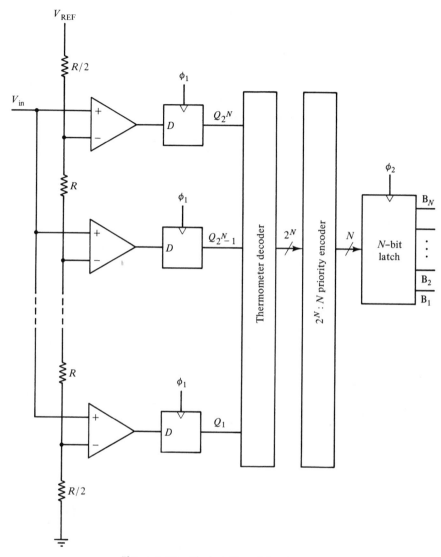

Figure 1-10   Flash A/D architecture.

The speed of the flash A/D is limited by the comparator switching speed and the propagation delay of the logic in the encoder. These two processes are often separated by the addition of a second stage of latches at the output of the encoder, which forms a pipelined architecture where two clock phases are required to produce an output. One sample is encoded into an N-bit result while the next sample is being quantized. Latches at the output suppress the transmission of transients that occur during the encoding process. In the timing diagram of Fig. 1-11, the analog input is sampled and the results are latched at the beginning of the first phase. While the encoding process proceeds, the comparators can resume tracking the input signal. In the second phase, the encoder output is latched and sent to the N-bit output.

Limitations in the resolution of the flash A/D result from the difficulty of integrating the 2^N comparison stages. Each additional bit of resolution causes a doubling of the amount of circuitry that is required, which will also tend to double the power dissipation of the A/D. The real situation can be even worse, since obtaining a twofold increase in resolution while keeping conversion speed constant will likely require even more power dissipation in the comparator. The matching and accuracy re-

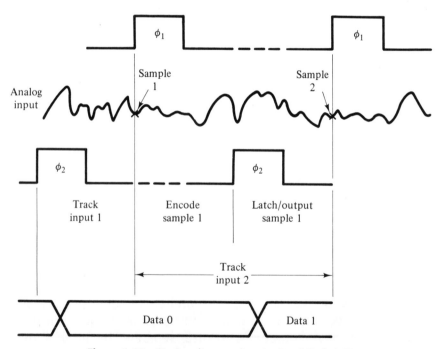

Figure 1-11   Timing diagram for pipelined flash A/D.

quirements in the reference circuit also double, which places a further practical constraint on the number of bits in flash A/Ds.

In the ideal case, the flash A/D can be used to digitize dynamic signals without an additional S/H. To be completely usable in this way, the bandwidth and settling response of the comparator must be sufficient to sample to the desired resolution signals which contain frequency components at least up to the Nyquist limit of one-half the sampling rate. In pulse digitizing applications, frequencies beyond this limit may be presented to the input. These requirements for dynamic performance, examined in greater detail in the following chapters, will be seen to form the most severe limitation on achievable resolution in flash A/Ds.

Type III: Subranging A/Ds

Subranging A/Ds, which are also known as two-step or two-pass A/Ds, combine some aspects of the pipelined A/D with those of the fully parallel A/D. Care must be taken to distinguish among these types of A/Ds, since aggressive marketing occasionally causes confusion by misappropriating the label of "flash A/D" for use with the subranging devices. A description of the typical architecture for this combination serial/parallel A/D, as shown in Fig. 1-12, will make the differences clear.

The first pass in a subranging conversion consists of a coarse quantization of the MSBs, digitizing the input signal to the resolu-

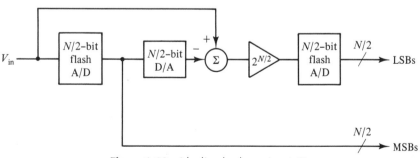

Figure 1-12   Idealized subranging A/D.

tion of the first flash A/D. In this simplified example, $N/2$ bits of the overall result are determined at this stage. An external S/H must precede the flash A/D, both to reduce the effects of internal timing errors and to hold the signal stable for the second-pass conversion.

The MSBs from the flash A/D are immediately reconverted to analog form in the $N/2$-bit D/A, which must maintain the accuracy of the signal to at least the desired number of bits in the overall converter. In this idealized example, the flash A/D is assumed to be a perfectly uniform quantizer, with all other error sources that affect accuracy and linearity set to zero. When the analog signal from the D/A is subtracted from the S/H output, the difference signal that is formed will then contain only the quantization error of the first conversion step. The magnitude of this error defines the end points of the "subrange," which remains to be quantized in the second A/D.

An ideal flash A/D, with uniform steps of 1 LSB, would have a worst-case error which is equivalent to a range of $2^{N/2}$ LSBs for the N-bit two-pass converter. One method that may be thought of for determining the $N/2$ LSBs is to directly digitize the error signal in a second flash A/D that spans this range. However, this would be impractical since the resolution of the second flash A/D would have to be $2^{N/2}$ times greater than what was required in the first A/D. Such a converter would be extremely difficult to implement for high-resolution applications.

A much better approach is to use an identical flash A/D in the second pass, possibly recirculating the error signal back to the same device. This would permit matching of the real error sources that affect the quantization process. An identical flash A/D can be used if, instead of reducing the reference voltage range, the error signal is multiplied by a factor of $2^{N/2}$ so that it spans a range equal to full scale. This process of expanding the subrange is illustrated in Fig. 1-13 for an ideal 4-bit converter, with two bits of digitization in each pass.

The 2-bit A/D in the example is capable of explicitly quantizing four distinct levels. As was shown in Fig. 1-10, the limit of the range may also be detected to provide the overflow signal. Each of the four identical coarse divisions that can be defined

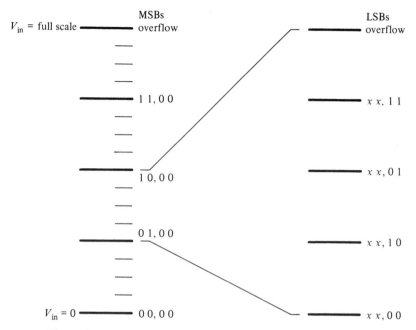

Figure 1-13   Expanding subrange for two-pass A/D converter.

in Fig. 1-13 spans four additional fine quantization levels. The two MSBs that are digitized determine in which coarse division, or subrange, the final result lies. This subrange is expanded by multiplying the error signal by four so that its amplitude can span up to the full-scale reference voltage range. The two bits of the second pass are then shifted two positions to the right in the 4-bit output, effectively dividing out the subrange gain factor that was used.

To assure overall performance, the gain of the subrange error amplifier combined with the D/A must not cause the second pass to overflow. For real flash A/Ds, the errors in first-pass quantization must also be considered. Subsequent chapters will describe these errors and the compensation techniques which must be used to properly implement subranging A/Ds.

2 ///////////////////

Looking Inside Flash A/D
Converter ICs

The rapid advances in the capabilities of flash A/D integrated circuits reflect the breakthroughs in semiconductor processing that have ushered in the VLSI era. Achieving integration levels of one million transistors on a single silicon chip, as in recent microprocessor and memory ICs, has required the shrinking of internal transistor dimensions to 1 μm (10^{-6} m) and below. At the same time, the addition of multiple interconnect layers has been required to provide high-speed routing channels for the increased number of signals. While being driven primarily by digital applications, these developments provide technologies that require little or no modification in order to be exploited in the analog domain. This has led to the development of monolithic, fully parallel 10-bit A/Ds and has extended conversion speeds to 200 million samples per second and beyond. Achievements in VLSI, rather than causing the digital replacement of analog functions as some had predicted, will result instead in even more sophisticated analog/digital ICs in the future.

CMOS (complementary metal oxide semiconductor) and high-speed bipolar processes, predominantly ECL (emitter-coupled logic), have both contributed to increases in circuit speed and complexity for flash A/Ds. Each technology has advantages for different applications, which has naturally resulted in the emergence of BiCMOS processes that exploit the best of both. For designers who are selecting and using flash A/Ds, it is important to understand that fundamental differences in performance and functionality exist between CMOS and bipolar/ECL devices. This chapter will review the design and operation of the key internal circuit blocks that make up each type of flash A/D converter. Since real devices never perform exactly as their ideal functional descriptions imply, the error sources that limit the performance of flash A/Ds are examined to form more complete and realistic models.

Design of CMOS Flash A/Ds

The CMOS Comparator

CMOS technology originally evolved as a low-power alternative to NMOS (*n*-type metal oxide semiconductor) processes for the fabrication of digital logic circuits. A shift to MOS technology occurred because the smaller dimensions and lower power of MOS transistors, compared to bipolar logic, became necessary to integrate early microprocessors and to increase the capacity of memory chips. It became evident very early, however, that MOS possessed unique characteristics for analog signal processing, resulting in the development of switched-capacitor filters, data converters, and coder–decoders (CODECs) for telecommunications.

Although the terminology remains the same, today's technology has all but eliminated the use of metal-oxide for the gate terminal of CMOS transistors, depending instead on polycrystalline silicon gates to push dimensions down to the 1-μm level. Other advances in process technology have greatly reduced the destructive latchup phenomenon that plagued earlier devices. Speed of CMOS circuits is also now competitive with bipolar, except for the more power-hungry ECL components.

All these advances have been put to good use in the switched-capacitor CMOS comparator, which is the most critical circuit element in the design of any flash A/D converter. Although other techniques can be used in CMOS flash A/Ds, the structure represented in Fig. 2-1 was the first and still is the most popular. Other

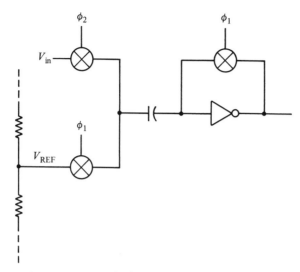

Figure 2-1 Switched-capacitor CMOS comparator.

approaches have far more in common with the popular bipolar circuits, which will be discussed later in this chapter.

The CMOS comparator consists of a pair of switches to alternately sample the input signal and a reference voltage level, an AC coupling capacitor, and an inverter to which feedback can be applied by a third analog switch. The feedback is necessary to bias the comparator and perform the autozeroing function, which is described below.

Figure 2-2 illustrates the DC transfer characteristic of a CMOS inverter. The IC designer typically seeks to match the gain of the PMOS and NMOS transistors in order to obtain equal drive capability for both the pullup and pulldown conditions. Because of inherent differences in physical characteristics of the two transistor types, judicious selection of each device's dimensions is required to obtain electrical matching. When this ideal is achieved, the logic transition voltage will occur exactly at one-half the V_{DD} supply voltage level. For digital circuits, this also has the advantage of maximizing noise immunity for both logic transitions.

The low power of CMOS logic results from the push-pull action of the complementary transistors. As Fig. 2-2 illustrates, there is

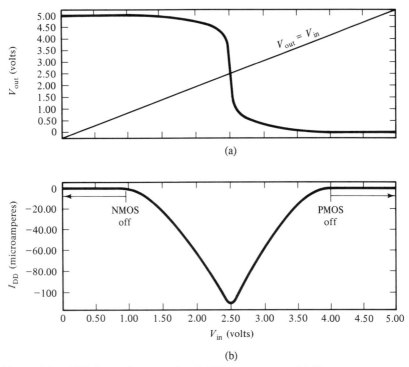

Figure 2-2 (a) DC transfer curve for CMOS comparator. (b) Bias current versus input voltage.

a very narrow range of operation where both the *n* and *p* transistors are conducting current. When a stable logic level is reached, only one transistor in the pair is on, resulting in zero current consumption from the supply. Since the output typically drives a similar high-impedance gate, which does not require DC bias current, the result is zero static power dissipation.

The narrow range where both transistors conduct in the CMOS inverter is exploited for the analog comparator. The slope of this transition region, dV_{out}/dV_{in}, determines the DC gain of the comparator. The purpose of the feedback switch is to force the comparator to a state where this high-gain condition exists. When the switch is turned on, a bias point is established where the 45° line of $V_{out} = V_{in}$ intersects the transfer characteristic of the inverter. In the ideal case, this point also occurs in the middle of the transition region.

One purpose served by the input capacitor is to hold the bias point on the input when the active feedback is removed by turning off the switch. However, Fig. 2-3 illustrates that in the non-ideal case transistor mismatch results in a DC offset of the bias point. Some mismatch is inevitable due to differences in the geometry and physical parameters of the p- and n-type transistors. Throughout an entire chip, each comparator may have a slightly different trip-point, which will also drift with changes in operating temperature. The accumulated offsets, if coupled directly to the input, would result in linearity errors in a flash A/D. It can be seen that the capacitor and feedback form an autozeroing function, allowing each comparator in the parallel array of a flash A/D to be set to its own unique bias point. As long as this point is maintained within the high-gain region of operation, accuracy is not affected. This is one of the key advantages of the CMOS architecture, making it insensitive to variations in operating temperature and device mismatch.

During the period that the comparator is autozeroed and storing the bias voltage on the summing node side of the capacitor, an

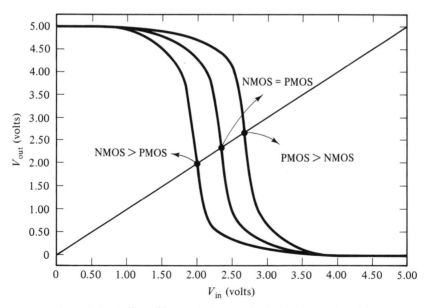

Figure 2-3   Offset of bias point versus NMOS/PMOS mismatch.

analog switch simultaneously charges the other capacitor terminal to a unique reference voltage that is established by a resistive voltage divider. In a CMOS process, the analog switch is capable of charging the capacitor with no loss due to series voltage drop. The final state of the comparator array at the end of the ϕ_1 interval can be summarized as follows: each input capacitor is charged to a voltage representing one of the 2^N quantization levels to which the input signal will be digitized while holding the comparator in a balanced condition ideally centered between the two logic levels. The amount of charge on each capacitor is different, and the polarity reverses from negative to positive over the span of the reference voltage; but the key element is that each comparator is balanced when the reference and feedback switches are turned off.

In ϕ_2 the input signal is sampled onto the capacitor while the comparator acts as an open loop amplifier. With a high input impedance on the summing node, charge on the capacitor will be preserved, causing the inverter input to change in voltage by the difference between the levels of the precharged reference and the input signal. Each comparator amplifies this difference signal with a gain equal to the slope at the autozero point in the transfer curve, establishing a logic 1 or logic 0 level at each output. Those comparators that see large difference signals will easily swing to one of the supply rails, either ground or V_{DD}. The key to a proper design is to assure that each comparator has sufficient gain so that a stable logic level can be established when the difference signal is at a minimum overdrive condition. This sensitivity, or noise threshold, must be at least equal to the expected resolution of the A/D, $\pm\frac{1}{2}$ LSB or $V_{REF}/2^{N+1}$.

CMOS A/Ds, which use a switched-capacitor sampling comparator, are often described as providing an inherent sample/hold function, eliminating the need for an external device. In the case of flash A/Ds with the operation outlined above, such a description is inaccurate, although an external S/H may still be unnecessary. Although the charge acquired from the reference voltage is indeed held on the input capacitor, during the interval when the capacitor is switched to the input signal the comparator more closely resembles a *track/hold*, since the summing node at the inverter input must accurately follow the AC component of the signal for the entire duration. It will be shown below that the bandwidth and slewing ability of the CMOS flash A/D is critically dependent on how well this function is performed.

Latch and Thermometer Decoder

At the end of the ϕ_2 tracking interval, the state of the comparators is frozen by turning off the input switches and storing the results in a bank of 2^N latches. It is the level of the input at the end of this interval that is finally quantized. A typical circuit for a single-stage CMOS latch is shown in Fig. 2-4. This circuit gates the comparator output to the first inverter's input, transmitting a valid logic level after the comparator has settled properly. During the autozeroing clock phase, the latch output is held by enabling the positive feedback to the input from the second inverter.

The digital output at this point should represent a thermometer code, as described in Chapter 1. A continuous string of logic 1s will be generated by the comparators which detected that the input was higher than their precharged reference voltage. This will be followed by a continuous string of 0s from the comparators that were precharged to a level greater than the final input voltage. The boundary between the 1s and 0s must be detected in order to establish the segment of the quantization range in which the input lies.

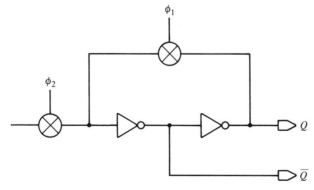

Figure 2-4 CMOS latch.

Figure 2-5 illustrates a portion of a thermometer decoder in the vicinity of the 1/0 boundary. In this example, the result should indicate that the input is greater than the third quantization level but not sufficient to set the fourth output high. The final result at the output should then be the N-bit code associated with the third

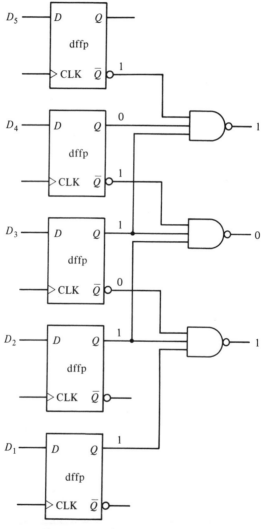

Figure 2-5   Thermometer decoder.

level. A 3-input NAND gate is used to detect the 1/0 boundary condition. An active low signal from the NAND gate thermometer decoder will be produced only by the gate associated with the third stage, while all the gates above and below will produce logic 1 signals. By using a 3-input rather than a 2-input gate to detect the 1/1/0 condition, an extra margin of noise immunity is obtained. If a spurious 1 is produced within the string of 0s, it is prevented from corrupting the encoder output because of the disabling signal that is produced from the stage below.

2^N:N Priority Encoder

Although it performs the function of a priority encoder, determining the position of the highest comparator to produce a logic 1 output, the block that reduces the thermometer code from 2^N logic signals to an N-bit word usually is implemented in a ROM (read-only memory) or PLA (programmable logic array). The NAND gates in the thermometer decoder, by producing just one active output signal, deliver a unique address to this block for each quantization level of the flash A/D. By "hard-wiring" each NAND output to one encoded N-bit word, the final result will represent the position of the highest reference quantization level that was exceeded by the input signal.

In CMOS there are a number of options which can be used to implement the encoder. Each technique produces an architecture which can be described as a "wired-OR" array. For each one of the N bits in the output digital word, there are an equal number of ls and 0s that are produced to encode all of the 2^N possible codes. Thus, 2^{N-1} quantization levels produce the same result for any single bit and can be connected in parallel or "ORed" together. Of course, as has been shown above, only one input can actively produce a 1 or 0 for any particular sample of the input. The example in Fig. 2-6 will clarify this point for a 3-bit example. Each column forms 1 bit in the encoded result, while each row is driven from the output of a thermometer decoder gate. A dot at the intersection of a row and column represents an OR input term for that output bit.

For low power considerations, the ORing operation can be performed by an array of complementary transistors. For each 0

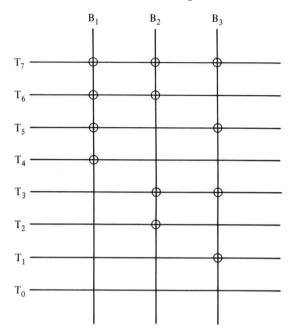

Figure 2-6   Wired-OR encoding of thermometer code.

that is produced, an NMOS device is used; while for each 1, a PMOS transistor is used. This approach is illustrated in Fig. 2-7. As for CMOS logic in general, current is drawn from the supplies only during the switching interval between logic levels. The disadvantage of this approach is that 2^N transistors are required for each bit (2^{N-1} NMOS and 2^{N-1} PMOS), and switching speed is limited by the parasitic capacitance each device adds to the output bit line.

The total number of transistors can be cut in half by using only NMOS devices to switch the output lines. As shown in Fig. 2-8, a pullup is required to produce a logic 1 output while the NMOS switches produce logic 0 signals. This technique will reduce the output logic swing and produce static power dissipation, since the current from the pullup transistor causes a voltage drop in the NMOS device for low output signals.

To eliminate static power while reducing the total number of transistors, a single clocked PMOS pullup can be used for each

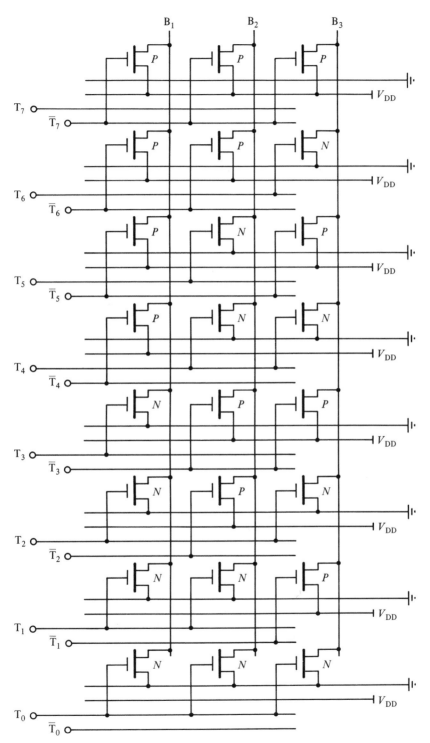

Figure 2-7 Full CMOS encoder.

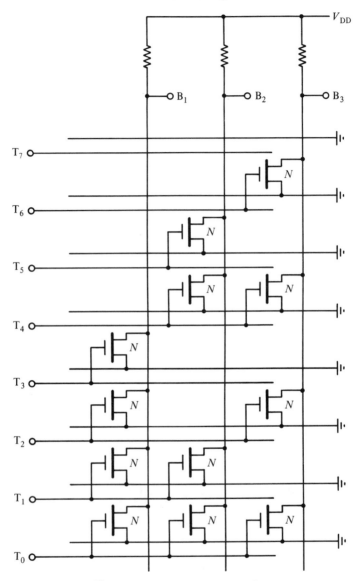

Figure 2-8   Static NMOS encoder.

bit line. This technique uses dynamic logic methods and requires precharging the bit lines to the high level prior to enabling the thermometer code to be presented to the encoder inputs. Speed of logic transitions is improved since the NMOS switches only have to discharge the capacitance on each bit line without actively pulling down the PMOS pullup, and logic 1 levels are set up before encoding is done.

Disadvantages of dynamic logic are inherent in this approach; high logic levels will decay, resulting in a minimum refresh rate. Low noise thresholds can also result. This condition is worsened because of the overall increase in digital noise that is generated from this approach. Observe that the bit line capacitance must return to a valid logic 1 during each clock period. This process occurs even if a string of consecutive 0s is ultimately produced, in effect transmitting a full amplitude signal at the sampling clock frequency on each bit line. Generating such signals on several bit lines simultaneously will generate considerable noise in the power supply and also result in large switching transients to ground when the bits are switched. Dissipation of the charging current will result in higher power than the complementary approach.

Design of Bipolar Flash A/Ds

The Bipolar Comparator

ECL logic evolved, and continues to be used, as the technology of choice when the fastest possible speed is required in a digital system. In such applications, the more popular TTL (transistor–transistor logic) bipolar technology could not keep up. The difference in performance came not so much from process technology innovations, which would benefit either logic form, but in the method used for creating the binary logic levels internally. TTL allowed transistors to switch fully from cutoff to saturation, while ECL transistors switch only between the active and cutoff modes. ECL eliminates charge storage delays that occur when switching bipolar transistors out of saturation, but it does this at the expense of higher continuous power dissipation. Speed increases are accompanied by reduced logic swings and the associated change in voltage supply levels to 0 and -5.2 V.

The first flash A/D ICs were implemented using the technology of the ECL gate families for speed considerations, but they generally maintained TTL compatibility at the interface to outside logic levels. As the switching speed of subsequent generations of TTL spinoffs increased, it became feasible to design bipolar flash A/Ds which do not require the addition of negative supplies to the systems in which they are used. The fastest bipolar flash A/Ds will continue to require ECL logic interfaces.

The core comparator of all bipolar flash A/Ds consists of a variation on the simple differential pair shown in Fig. 2-9. During the input sampling interval, the clock enables the emitter tail current source to flow through the pair of transistors Q_1 and Q_2, which will amplify the difference in voltage between the input signal and reference voltage level.

A DC transfer curve of the differential output voltage versus the differential input voltage is shown in Fig. 2-10a. In the ideal

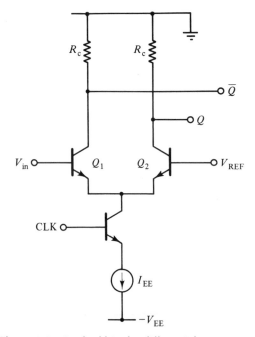

Figure 2-9   Strobed bipolar differential comparator.

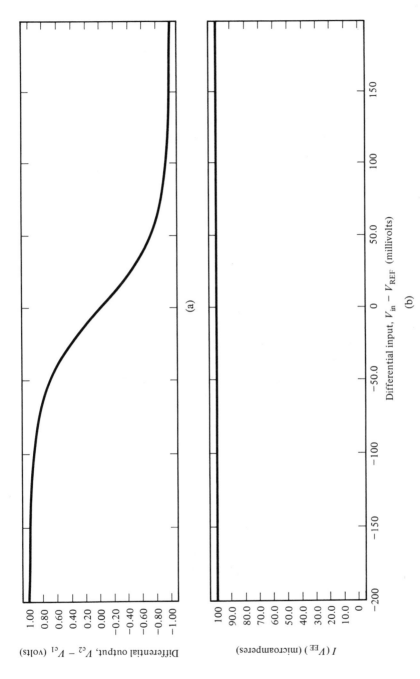

Figure 2-10 (a) DC transfer curve for bipolar differential comparator. (b) Bias current.

case when the analog input is exactly equal to the reference tap
voltage, the bias current splits equally in Q_1 and Q_2, resulting in
the differential output of zero. Small deviations from the reference voltage will be amplified, eventually limiting the output
when the total emitter current flows in one side of the pair. This
condition causes the differential output to switch to a value of
$\pm \alpha_F \cdot I_{EE} \cdot R_c$, where α_F is related to the forward transistor gain
by $\beta_F/(\beta_F + 1)$. An important point to observe from Fig. 2-10b is
that all of the bipolar differential comparators in a flash A/D
consume the same continuous power regardless of the state they
are switched to. This greatly contributes to the generally higher
power dissipation seen in such devices compared to their CMOS
counterparts.

A review of the simplified model for the bipolar differential pair
will illustrate some of the constraints that the circuit designer
must deal with. This is the well-known hybrid-π model, as shown
in Fig. 2-11. The intrinsic small-signal base-emitter resistance is
represented by r_π, and for this discussion it is assumed that the
output resistance is infinite and there is zero series parasitic base
and collector resistance. The emitter resistance R_{EE} is typically

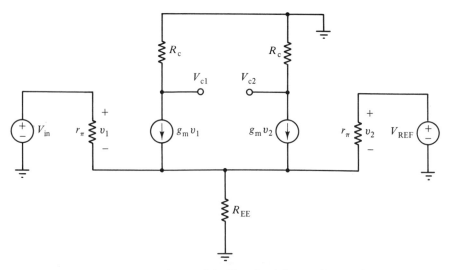

Figure 2-11   Hybrid-π model of bipolar differential comparator.

the output impedance of a current source transistor. The differential gain of this circuit is defined as $(V_{c1} - V_{c2}) / (V_{in} - V_{REF})$. A complete analysis of the differential pair is left to the reader as an exercise (it is covered in detail in many circuit design textbooks), but the final result shows that the differential gain reduces to just $-g_m \cdot R_c$. The transconductance g_m of the bipolar transistor is $I_c/V_t \approx I_{EE}/V_t$. The factor V_t is referred to as the thermal voltage and is typically 26 mV at room temperature. For A/D converters it is most critical to recognize that the required resolution, which dictates the voltage gain or sensitivity of the comparator, also directly determines the continuous power ($I_{EE} \cdot V_{EE}$) which is dissipated from the bias current sources in the differential pair.

The switching characteristics of the bipolar differential comparator can cause problems for both the reference and input signals that must drive the pair of transistors in flash A/D applications. Because the signals are DC coupled, a base current must be supplied from the reference resistor ladder to all the comparators that are above the level of the input signal. Some of the comparators may remain in the forward-active mode under the following condition:

$$V_{REF}(x) < V_{c2} = -I_c \cdot R_c$$

so that the required I_B is $I_E/(\beta_F - 1)$. The input resistance r_π, which is equal to β_F/g_m, represents a load which can only be increased by reducing the comparator gain ($g_m \cdot R_c$). What makes this situation worse, however, is that in the comparators with higher reference levels the input NPN transistor may become saturated, resulting in base currents which increase exponentially as the base-collector junction diode becomes more strongly forward biased. The resulting voltage drops in the reference resistor ladder, which will be a function of the input signal, would cause nonlinearities in the A/Ds transfer function.

For the input signal, which must drive the 2^N comparator inputs in parallel, a signal level dependent input impedance would result. At signal levels near the bottom of the reference ladder (REF−), a small number of transistors are active, resulting in a relatively high input resistance. As the input level increases, more transistors turn on, with some potentially reaching saturation. By drawing progressively larger currents from the input buffer amplifier, signal distortion would result.

Because of the variable input impedance, the differential pair in bipolar flash A/Ds is usually preceded by an emitter follower to buffer both inputs, as shown in Fig. 2-12. Although this technique does not eliminate the signal-level dependence of the comparator input impedance, it does increase the overall impedance by multiplying the input resistance of each stage by the factor of $(\beta_F + 1)$, which is a characteristic of the emitter follower configuration. The reference ladder resistance of bipolar flash A/Ds is typically kept on the order of 100–200 Ω, which further reduces the effect of base current drain.

Due to the direct coupling of reference and signal input signals, an additional limitation of the bipolar differential comparator is that DC offset voltages due to device mismatches in the four input transistors will directly affect the accuracy and incremental uniformity of reference tap voltages in the A/D. This error source results in nonlinearities which, as will be shown in Chapter 3, increase the total noise level of the converter.

An additional stage of gain is often provided ahead of the

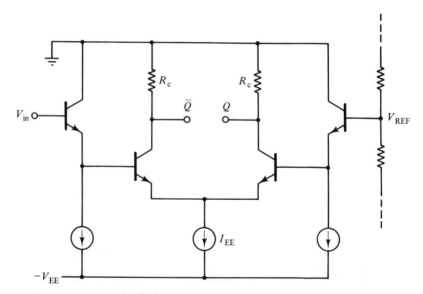

Figure 2-12   Bipolar flash A/D comparator with emitter-follower buffers.

latching comparator, especially in A/Ds with greater than 100 MHz of conversion speed. This stage operates purely as a differential amplifier to boost the overdrive of the comparator stage at high clock frequencies, thus reducing settling time. In such applications, the DC offset characteristics are again important in determining accuracy. The frequency response of the differential amplifier will directly determine input bandwidth limitations, since gain rolloff at higher frequencies attenuates the signal that is transmitted to the comparator input.

Comparator Latch

One common method of implementing the comparator latch is to cross-connect a second pair of transistors to the collectors of Q_1 and Q_2, as in Fig. 2-13. When the clock switches to ϕ_2, the emitter current is steered to these transistors where positive feedback results in a latch function, disabling the input signal from affecting the differential output voltage level. An advantage of this current steering technique is that no additional bias current is necessary to implement the latch function. However, in the latched mode the input transistor pair is cut off, which presents a sudden change in input impedance of the comparator. This can result in the "kickback" of clock noise to the input and reference, which can corrupt the conversion process.

The differential output of the first comparator is often fed to a similar second stage latch in a master/slave function. A second stage helps to reduce metastable (or indeterminate) states by providing additional amplification of the difference signal, and it extends conversion speed by holding a stable comparison result to the encoder for a complete clock cycle.

Thermometer and N-Bit Encoder

Thermometer decoding can be done either preceding or following the latch, and it is sometimes combined with the second latch stage, using ECL logic to compare one or two adjacent states. A 3-input OR gate, which can be used for this purpose, is shown in Fig. 2-14. This circuit employs three parallel input transistors, which are emitter followers from the respective comparator out-

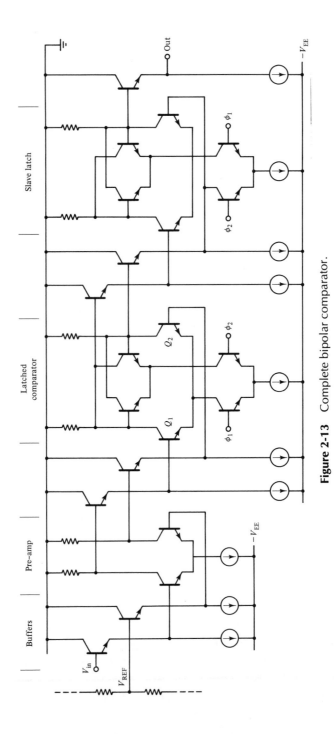

Figure 2-13 Complete bipolar comparator.

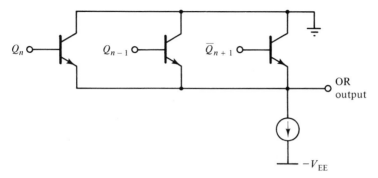

Figure 2-14 Three-input OR gate thermometer decode.

puts, to form a wired-OR connection. An AND function can also be created by stacking input transistors in series.

The latched thermometer code in a bipolar flash A/D is encoded into an N-bit word by a wired-OR array of NPN emitter followers. This is illustrated in Fig. 2-15 for a section of the three LSBs. With only one active high input from the thermometer decode, an emitter connection is required to pull up a bit line to a high level. While the encoder is often split into four stages, which are then ORed together, exactly half of the comparators within any group will produce pullup connections. The capacitance of the large number of parallel transistors will limit the switching speed, which will not generally be symmetrical since the 1–0 transition depends on pulldown from the output load circuit, which is generally a constant current source.

Error Sources in Flash A/Ds

Reference Circuit Nonidealities

A primary factor determining the basic DC linearity of any flash A/D is the matching that is obtained in the elements of the resistor divider. In the case of CMOS, precise setup of the capacitor levels during the autozero interval requires that each reference tap be equal and increase incrementally by an amount equal to $1/2^N$ times the full-scale reference voltage. In bipolar devices, uniform reference levels must be established without being af-

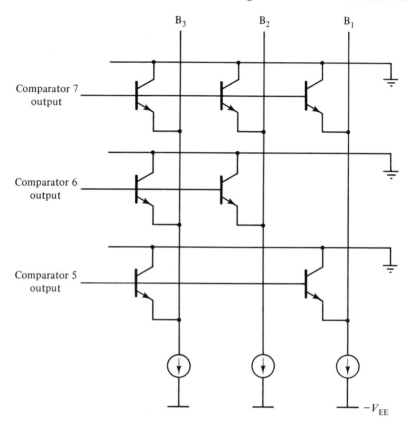

Figure 2-15   NPN wired-OR encoder.

fected by the base drive requirements of the comparator inputs. The accuracy with which this is accomplished depends partially on the patterning techniques that are used in the fabrication process. This is one example where the advances in VLSI technology can provide a spinoff benefit for analog circuits. Tighter dimensional control is required to reliably shrink integrated circuit features down to submicron levels. Even though the resistor element may not require the minimum allowable dimensions from a VLSI process, the inherent accuracy in patterning improves the overall matching.

Matching of the resistor ladder elements is also dependent on geometry, which is dictated by the IC layout and the materials that are available to the designer in any particular process. After determining the total resistor value, from speed and power considerations, the key parameter for the designer is the resistivity of the materials used to form the various conductive layers of the IC. Measured in ohms/square, this factor determines the length-to-width ratio or number of "squares" required to achieve a given value. Figure 2-16 illustrates the geometry of three identically valued resistors built with different materials that may be found in CMOS and bipolar processes.

Resistor values for each reference tap are typically required to be 1–3 Ω in order to maintain performance at flash converter speeds. As Fig. 2-16 shows, low resistivity materials, such as those used for interconnect metallization, are best suited to this purpose. The aspect ratio that results from using higher resistivity materials, such as the polysilicon material used for MOS gates and emitters in some bipolar processes, is impractical and would force the use of higher valued resistors and, consequently, result in lower speed. The geometry must be chosen carefully not only

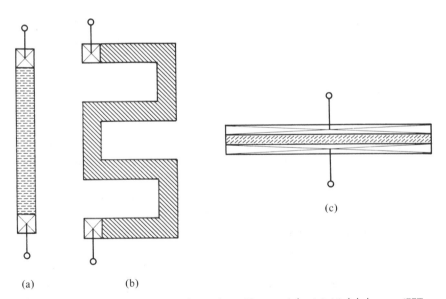

(a)   (b)

Figure 2-16 Resistor geometry in various IC materials: (a) Molybdenum (FET gates and interconnect), (b) aluminum (interconnect), and (c) polysilicon (FET gates).

for matching considerations but also to insure that the current carrying capacity of the metal conductor is never exceeded.

When reviewing manufacturers' data sheets, examination of the reference ladder resistance can provide a clue to dynamic performance as well as static power dissipation. Low resistor values are required in CMOS in order to recharge the input capacitors quickly after full-scale signal swings. In bipolar A/Ds, low resistance is required to suppress AC signals flowing in the ladder as a result of high frequency input signals, which cause the load to become dynamic as varying numbers of reference transistors in the differential pairs are cut off or turned on.

Charge-Feedthrough Errors

The description earlier in this chapter of the autozeroing technique for CMOS comparators was idealized in that it did not take into consideration transient effects which occur. Analog MOS switches, such as are used for the reference and feedback switches, inevitably develop undesirable parasitic capacitances between their gate and source/drain terminals. When the MOS switches are turned off, transient currents which flow through the parasitic capacitors can alter the charge which is finally stored on the input coupling capacitor of the comparator. Figure 2-17 schematically illustrates the location of these parasitic elements for NMOS switches, coupling into the summing node and reference input of each stage.

As an example of charge feedthrough effects in CMOS A/Ds, assume that a comparator at the bottom of the ladder has charged the input capacitor to near 0 V and that the summing node is balanced at 2.5 V. The gate terminals of the NMOS devices are at 5 V for the "on" condition. It becomes apparent that the terminal voltages of the parasitics on either side of the input capacitor are not equal. This will result in an unequal amount of charge transfer when the switches turn off, even if the two transistors are identical. The net effect is as if a signal voltage, either above or below the reference level, was transmitted to the comparator input. This will disturb the balanced condition, and this differential

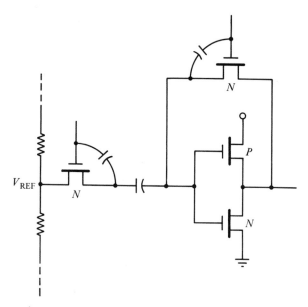

Figure 2-17   Parasitic capacitors in CMOS switches.

signal will be amplified by the comparator, resulting in false logic level output before any signal is applied.

One result of charge feedthrough is an offset voltage that is present at the comparator input at the beginning of the sampling interval. The input signal must overcome this error to get the comparator to switch states. This will partially explain why offset specifications are still included on CMOS flash A/D data sheets, even though the comparators are autozeroed to eliminate other DC errors.

Since the voltage across the parasitic capacitor from the reference sampling switch changes linearly over the range of the A/D, the effect on each comparator will be different. If the magnitude of the parasitic capacitor stays constant, then a linear change in feedthrough error only contributes to a gain error for the overall converter. However, the parasitic capacitors may have a voltage coefficient which causes the feedthrough errors to change in a nonlinear fashion. When this occurs, integral linearity is also degraded. A detailed description of these performance specifications is given in Chapter 3.

Input Capacitance and Settling Time

CMOS A/Ds

The input sampling capacitor of the CMOS comparator, which proves beneficial for DC performance, can cause particular problems for dynamic performance. The capacitance of the analog input is a limitation in any flash A/D, bipolar or CMOS, because of the load it presents to buffer amplifiers that must be used to transmit the signal that will be digitized. This capacitance will affect bandwidth, settling time, and slewing ability.

In the CMOS A/D, the fixed input capacitance is switched from the reference to the input. During the reference sampling interval, the input buffer drives a very high impedance from the switches in their off state (essentially an open circuit) and a small amount of background parasitic capacitance that inevitably results from the package pin to chip interface and internal signal routing. When the coupling capacitors are switched to the input, they instantaneously change the capacitive load on the buffer amplifier. What makes this situation especially troublesome is that at the moment of switching, the 2^N capacitors are initially charged to 2^N different voltages, with levels that differ incrementally from the signal by as much as $\pm V_{REF}$. With a low output impedance buffer, the input switching causes large transient current spikes to occur, which vary in magnitude and direction depending on the level of the signal and net charge on the array of capacitors.

The charge differential between the precharged levels and the amount required to establish the input signal must be dissipated on each element of the capacitor array in a settling time sufficient to maintain the desired conversion rate and accuracy. When the input signal is near ground, each capacitor initially stores a voltage which is greater than the input level. The net positive differential causes a positive current spike at the beginning of the sampling interval, as shown in Fig. 2-18a. When the signal is near the midpoint of the conversion range, there are as many capacitors charged to a voltage above the input level as there are capacitors charged below. The net charge cancels, resulting in a minimal spike, as shown in Fig. 2-18b. Likewise, when the signal is near the top end of the voltage range a net negative charge differential causes the spike to appear as in Fig. 2-18c.

The worst-case situations occur when the input signal is near the limits of the conversion range, either zero or V_{REF}. A precise analysis of the transient response in the reference network requires an extensive computer simulation. However, for purposes of explanation, the two simultaneous effects which occur to bring the capacitors to the input potential can be analyzed separately. The first effect is the contribution from charge redistribution. As Fig. 2-19 shows, the parallel connection of two identical capacitors with initially different potentials will ultimately result in a potential equal to one-half the total. Charge conservation requires that the total charge remain constant, so the potential difference causes charge to redistribute from $C1$ to $C2$ to equilibrate the voltage:

$$Q_T(0) = C \cdot (V_{C1} + V_{C2})$$

which, after switching, becomes

$$V_{CT} = \frac{Q_T(0)}{2C} = \frac{V_{C1} + V_{C2}}{2}$$

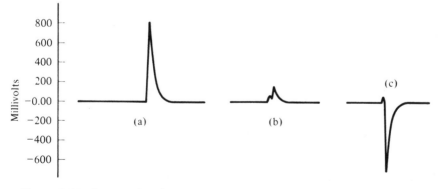

Figure 2-18 Input spikes from CMOS switched-capacitor input. (a) $V_{in} = 0$, (b) $V_{in} = V_{REF/2}$, and (c) $V_{in} = V_{REF}$.

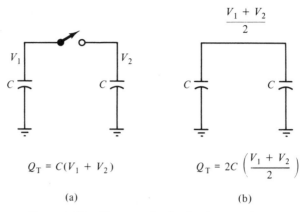

$$Q_T = C(V_1 + V_2)$$

$$Q_T = 2C\left(\frac{V_1 + V_2}{2}\right)$$

(a)   (b)

Figure 2-19   Charge redistribution in capacitors.

At the end of the reference charging interval, the array of capacitors can be modeled in exactly the same way as this simple example. The array can be seen to have odd symmetry through its midpoint, as illustrated in Fig. 2-20. The two capacitors immediately above and below $V_{REF}/2$ have potentials of $V_{REF}/2 - \frac{1}{2}$ LSB and $V_{REF}/2 + \frac{1}{2}$ LSB, respectively. There are similarly matched pairs across the entire array. If there were no input buffer present when the closing of the input sampling switches connects all the capacitors in parallel, complementary capacitor pairs would cause charge redistribution to occur until a final potential of $V_{REF}/2$ was achieved. This condition can be used to approximate the additional charging current that will be required from the buffer to make up the difference between the signal level and $V_{REF}/2$.

Assume an input signal at 0 V, with a requirement for settling to $\frac{1}{2}$ LSB within the sampling interval. For the purposes of this model, the on-resistance of the analog switches can be lumped with the output impedance of the buffer amplifier so that an equivalent circuit, as in Fig. 2-21, may be derived. The settling performance of the simple RC network is described by the following equations.

$$V_{in}(T_s) = \frac{V_{REF}}{2} \cdot e^{-T_s/R_sC_T} = \frac{V_{REF}}{2^{N+1}}$$
$$T_s = R_s \cdot C_T \cdot \ln(2^N)$$
$$T_s = 0.69 \cdot N \cdot R_s \cdot C_T$$

As an example for an 8-bit A/D, with typical numbers for R_s of 100 Ω and effective input capacitance from C_T of 30 pF, the settling time to $\frac{1}{2}$ LSB would be 16.5 nsec. This equation represents a fundamental limit on the speed-resolution product of an autozeroed switched capacitor CMOS flash A/D.

Other considerations besides settling time, such as matching requirements and minimization of charge feedthrough effects, determine the minimum input capacitance that can be used. The total value will be at least double with each additional bit in the conversion, assuming that adequate resolution can be obtained from the comparator without increasing the capacitance further. Realizable buffer amplifiers cannot achieve zero impedance under dynamic conditions, and the large surge currents which would flow from the output transistors at the switching instant must be limited in any case. Other parasitic effects, such as wire inductance, eventually limit high slew rate signals. With a fully parallel approach based on the switched capacitor technique,

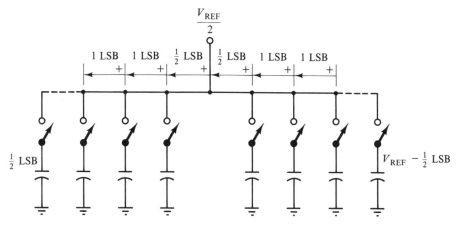

Figure 2-20 Odd symmetry in CMOS capacitor array after autozeroing.

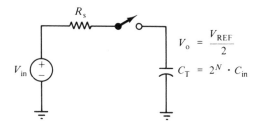

$$V_o = \frac{V_{REF}}{2}$$

$$C_T = 2^N \cdot C_{in}$$

Figure 2-21   CMOS settling time equivalent circuit.

maximum conversion rate will be limited to about 30 MHz in a true 8-bit A/D. Higher resolution cannot be obtained at the same speed without nonparallel architectures or the use of different comparator techniques.

Bipolar A/Ds

In bipolar A/Ds the capacitors that limit dynamic performance are undesirable parasitic elements that are an intrinsic component of all $p-n$ semiconductor junctions, such as those which form the transistor terminals. An important difference between sampling capacitors of CMOS A/Ds and junction capacitors is that the value of junction capacitors is determined from a non-linear function of voltage and temperature. This is an effect which is exploited in circuits which use varactor diodes as a variable tuning element. In order to emphasize this point, the equation from semiconductor physics which describes the capacitance per unit area of a reverse biased junction is given below:

$$C_j = \left[\frac{q \cdot \varepsilon_{si} \cdot N_p \cdot N_n}{2 \cdot (N_p + N_n) \cdot (V_{bi} + V_r)} \right]^{1/2}$$

In the equation q refers to the electron charge, ε_{si} is a material constant representing the permittivity of silicon, and N_p and N_n represent the doping concentration of ions in the p-type and n-type regions, respectively. The reverse-bias potential is represented by V_r, and V_{bi} is known as the built-in potential of the junction, which is given by the following:

$$V_{bi} = V_t \cdot \ln \frac{N_p \cdot N_n}{n_i^2}$$

The thermal voltage V_t is directly proportional to temperature, and n_i is the intrinsic concentration of carriers in the material before doping is used to form the junctions. The capacitance equation can be greatly simplified by combining the constant terms

$$C_j = \frac{C_{j0}}{\sqrt{(V_{bi}(T) + V_r)}}$$

Thus the parasitic junction capacitance follows an inverse square-root relationship with both temperature and reverse-bias potential. Empirically, the exponent of the denominator may deviate from a square root due to nonuniform doping of the $p-n$ regions over their contacting areas. However, the nonlinear characteristic of the relationship remains as illustrated in Fig. 2-22.

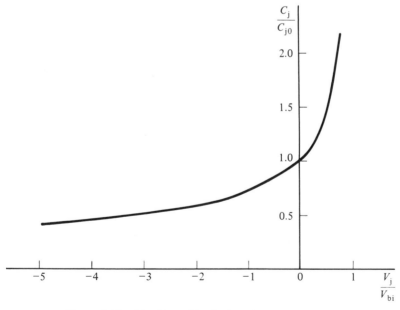

Figure 2-22 Nonlinear bipolar junction capacitance.

The effect of the nonlinear capacitance in a bipolar flash A/D is a complex one. The base-collector junctions of the input emitter followers remain reverse biased under all conditions of normal operation. This will contribute a signal level dependent capacitance, as shown in Fig. 2-22, which increases as signals become more positive. The base-emitter junction is normally forward biased, and rather than increasing to infinity as V_{bi} is approached, it will contribute a capacitance approximately equal to the value of C_j just prior to turn on. Worst-case settling time for a buffer amplifier to drive the input will occur as signals approach the positive reference potential, where parasitic capacitance is a maximum.

Input Impedance and Frequency Response

CMOS A/Ds: Input Resistance

Some models that have been proposed for the input impedance of the CMOS flash A/D utilize a standard equivalent circuit that is derived from the underlying principle of switched-capacitor filter design. These models replace the switched capacitor with a resistor that has a value which would develop an equivalent average current flow over the sampling clock period. The governing equation for this model is

$$R_{sc} = \frac{1}{F_c C}$$

The switched-capacitor resistor has a value which depends only on the clock sampling rate (F_s) and the value of the capacitance. This model is fine for applications with a high degree of oversampling (i.e., many times greater than the highest signal frequency) but is inappropriate for the case of flash A/D conversion where instantaneous response is important. As was described above, the input impedance of the CMOS flash A/D is complex, resulting from the three distinct states of autozeroing, charge redistribution, and input sampling. It is the characteristic of the A/D during the latter two states that most determines the drive requirements for external buffer amplifiers. Settling time requirements (described above) show that during the switching transient the input can either sink or source current depending on the level of the input signal. The same simple RC network will describe the impedance characteristic during the rest of the sampling interval.

With a low voltage coefficient on the capacitors, there are no nonlinear effects to distort the signal that is presented to the comparator input. It is important that any voltage coefficient in the coupling capacitor is minimized, so that gain is constant over the array of comparators and does not degrade dynamic linearity. This results in a load to the buffer amplifier which does not change with the DC level of the signal, enhancing stability.

Bipolar A/Ds: Input Resistance

As was described previously, all bipolar flash A/Ds precede the differential comparator inputs with emitter followers to buffer the dynamic load. Like the CMOS A/D, this load will change with the phase of the sampling clock, and it will also be dependent on signal level. The buffer provides an input which multiplies the load resistance seen at the emitter by the factor of $\beta_F + 1$. Figure 2-23 illustrates the three possible conditions which may exist simultaneously at various points in the bipolar A/D during a sampling interval.

The condition represented in Fig. 2-23a occurs during the latched mode phase or wherever the input transistor is cut off by a signal level below the reference level. For such comparators the load of the emitter follower, which may be a resistor or current source, results in a relatively high input resistance. Input current will be only $I_L/(\beta + 1)$ or $(V_{in} - V_{BE})/[(\beta + 1) \cdot R_L]$.

In Fig. 2-23b the case of transistors in the forward-active mode of operation is represented. As the reference level is exceeded the input to the differential pair turns on while the reference transistor eventually switches off. The impedance seen at the input is still relatively high, with an additional current of $I_{EE}/[\beta + 1]^2$ being drawn for the base current of the comparator.

The comparators in the active mode will have collector terminal voltages of $-\beta \cdot I_{EE} \cdot R_c/(\beta + 1)$. With the assumption of

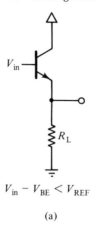

$$V_{in} - V_{BE} < V_{REF}$$

(a)

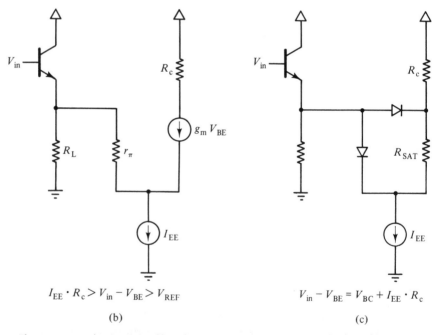

$$I_{EE} \cdot R_c > V_{in} - V_{BE} > V_{REF}$$

(b)

$$V_{in} - V_{BE} = V_{BC} + I_{EE} \cdot R_c$$

(c)

Figure 2-23 Three states of bipolar comparator input: (a) Latched mode or input cutoff, (b) active mode, and (c) saturated mode.

$\beta \gg 1$, at increasing signal levels the input transistor will approach saturation at the point where

$$V_{in} - V_{BE} > -(I_{EE} \cdot R_c)$$

Eventually, the input voltage can be sufficient to forward bias the base-collector junction diode. If this occurs, the effective β of the transistor drops, causing an increase in the base current that is drawn from the input. As shown in Fig. 2-23c, for large overdrive conditions the load impedance at the emitter follower output decreases, causing input current to increase to $(V_{in} - V_{BE})/[(\beta + 1) \cdot R_{eff}]$.

CMOS A/Ds: Small-Signal Bandwidth

The first factor that must be considered in determining the bandwidth of the CMOS flash A/D is the effect on the input buffer amplifier from the converter's capacitive load. The circuit of Fig. 2-21, which was used to determine settling time limitations, also holds for determination of bandwidth. The key simplifying assumption which is made here is that the input load of the A/D, in combination with the buffer output resistance, creates the dominant pole in the amplifier's transfer characteristic. This approximation is adequate, since the bandwidth of 100 to 150 MHz which can be expected from devices capable of meeting the drive requirements is much greater than the conversion speed of CMOS A/Ds. Again using typical parameters of 100 Ω for the source impedance and 30 pF for the total input capacitance from all of the A/D's comparators, the 3-dB bandwidth of the single-pole circuit is

$$f_{3\,dB} = \frac{1}{2\pi RC} = 53 \text{ MHz}$$

Internal to the A/D, the small-signal bandwidth of each comparator is dependent on the individual transistor and capacitor parameters chosen by the designer. A model for the circuit is shown in Fig. 2-24a. The equivalent circuit consists of the AC coupling input capacitor and a small signal model for the MOSFET transistors. A voltage-dependent current source models the

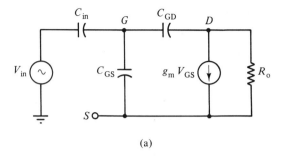

(a)

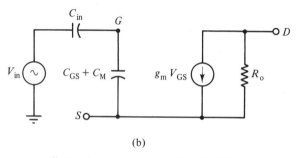

(b)

Figure 2-24   (a) Small-signal equivalent circuit for CMOS comparator. (b) Simplified model with Miller capacitance.

transconductance, while the resistor models the output impedance. The transconductance g_m depends on the fabrication process and resulting physical parameters for a given bias current, but it is always smaller for PMOS compared to NMOS transistors. By making the PMOS transistor physically larger than the NMOS transistor, their respective gains can be made approximately equal. The output resistance R_o is determined by the bias current, which is chosen by the designer to meet power dissipation requirements.

The circuit of Fig. 2-24a can be reduced to the effective circuit in Fig. 2-24b by taking into account the Miller effect. This rule states that the parasitic capacitance from input to output of the comparator, which is C_{GD} in this case, has the same effect on performance as a similar component placed at the input with a value multiplied by the gain as follows:

$$C_M = (1 + A_v) \cdot C_{GD}$$

For the CMOS comparator, the low-frequency voltage gain is simply $g_m \cdot R_o$ for the composite of the PMOS and NMOS transistors. Note that at the input, parasitic capacitance will create a voltage divider which reduces the gain of the comparator by a constant amount but does not directly affect its bandwidth:

$$\frac{V_{GS}}{V_{in}} = \frac{C_{in}}{C_{in} + C_{GS} + (1 + A_v) \cdot C_{GD}}$$

$$\approx \frac{1}{1 + (1 + g_m \cdot R_o) \cdot C_{GD}/C_{in}}$$

An important difference from bipolar flash converters, which are described in the next section, is that the attenuation of the AC-coupled signal being digitized does not necessarily result in loss of accuracy. The comparator gain is designed to achieve the required performance, including the attenuation at the input.

In actual application, the frequency response of the CMOS comparator will be determined by the load capacitance at its output. This results from driving the input of subsequent stages as well as parasitic junction capacitance. When these conditions are taken into account the gain of the circuit, using the variable S from Laplace transform analysis, is

$$\frac{V_o}{V_{GS}} = \frac{-g_m \cdot R_o}{(1 + R_o \cdot C_L \cdot S)}$$

The low frequency gain of the circuit rolls off at higher frequency with a 3-dB point at $1/(2\pi R_o C_o)$. However, the critical question to be answered in this analysis is what gain bandwidth is required to maintain the resolution of the A/D at high frequencies. The requirement from the comparator output is for a signal of sufficient magnitude to properly drive the latch to a valid logic level. A 3-dB reduction in output amplitude for the CMOS comparator may still provide enough gain to drive the latch without a significant loss of resolution. Because of the AC-coupled comparator, deteriorating frequency response will not cause the A/D to exhibit compression of the dynamic range at the output. Autozeroing keeps all of the comparators active in the quantization process.

Effective resolution will eventually degrade at higher frequency as the attenuated internal signals become affected by a loss in signal-to-noise ratio. However, the extent of this degradation cannot be predicted directly from parameters such as 3-dB bandwidth without knowing the sensitivity of the latch. In cases where sufficient gain bandwidth cannot be provided by a single stage, adding a second AC-coupled stage will improve performance. Figure 2-25 presents a Bode plot of AC performance from a typical CMOS comparator.

Bipolar A/Ds: Small-Signal Bandwidth

The intrinsic capacitance of the junction diodes in the bipolar transistor, which was described earlier, affects the frequency response of the input buffer amplifier in much the same way as the CMOS example described above. A major difference is the

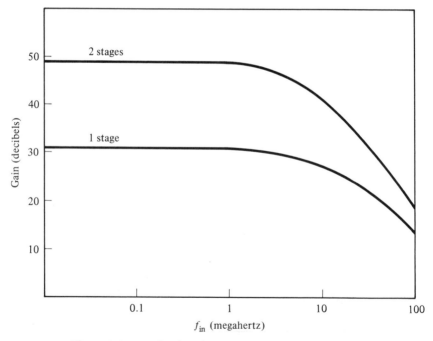

Figure 2-25   Bode plot of a two-stage CMOS comparator.

voltage coefficient of the reverse-biased base-collecter junction capacitance, which will cause greater attenuation as the DC level of the signal increases toward the collector supply voltage. Figure 2-26 shows the equivalent circuit for the emitter follower input to bipolar flash A/Ds. The base-emitter and base-collector junction capacitors are represented by C_π and C_μ respectively. The physical size of the transistor and the particular doping levels of the fabrication process will determine the magnitude of the junction capacitance. Minimum sizes are set by the patterning capability of the process and matching considerations which determine DC linearity errors. Typical values for the input capacitance in an 8-bit A/D of 30 pF or more, as used in the CMOS example, are common.

Bipolar flash A/Ds perform differently than CMOS A/Ds because the emitter followers act as internal buffers which must directly transmit the input signal to the comparator. The frequency response of this signal path is the primary limitation on dynamic performance of the A/D. High-frequency signals will be preattenuated at the comparator input, causing a compression of dynamic range, which can be directly observed as a reduction in the peak magnitude of the digital output code.

Figure 2-27 illustrates the small-signal equivalent circuit for the signal path to the differential comparator input, including parasitic capacitors. The circuit is dominated by the large Miller capacitance at the input to the comparator. Miller capacitance results whenever an amplifier has capacitance between its input and output, which in this case is the base-collector junction of the comparator input. The equation for this capacitance is

$$C_M = (1 + g_m \cdot R_c) \cdot C_\mu$$

where g_m is set by I_c/V_t, as discussed earlier, and C_μ is the base-collector capacitance. Setting the DC gain ($g_m \cdot R_c$) to provide the required resolution also strongly determines the comparator's input bandwidth.

In the equivalent circuit of Fig. 2-27, R_{in} combines the intrinsic base resistance of the comparator input with the output impedance of the emitter follower which drives it. The 3-dB bandwidth of this circuit is

$$f_{-3\text{ dB}} = \frac{1}{2\pi \cdot R_{\text{eq}} \cdot C_{\text{T}}}$$

where

$$R_{\text{eq}} = R_{\text{in}} \| r_\pi$$

and

$$C_{\text{T}} = (1+A_v) \cdot C_\mu + C_\pi$$

As an example, with $R_{\text{eq}} = 500$, $C_{\text{T}} = 20$ pF, and gain of 100 for the comparator, the 3-dB frequency is 15.9 MHz. Because of the limitations caused by the Miller capacitance in the differential comparator, bipolar flash A/Ds often have input bandwidth that is less than that of CMOS A/Ds with similar conversion speed.

Slew-Rate Limitations and Large-Signal Bandwidth

CMOS A/Ds

Because the CMOS comparator is autozeroed on each conversion cycle, input overload and saturation effects which occur in bipolar comparators are not relevant. However, the large-signal bandwidth will be dramatically affected if the slewing threshold is exceeded.

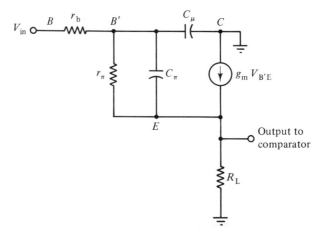

Figure 2-26 Equivalent circuit for input to bipolar flash A/D, including capacitors.

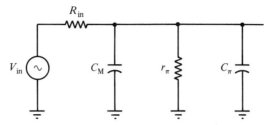

Figure 2-27   Small-signal equivalent circuit for input to bipolar comparator.

To see how this threshold is determined, Fig. 2-28 again illustrates the schematic of the CMOS autozeroed comparator, but it also includes parasitic diodes, which are inherent elements resulting from the fabrication process. The MOSFET device source and drain terminals are formed by creating a p–n junction with the opposite polarity material which forms the substrate and backgate terminal. These diodes are normally maintained in a reverse-biased condition by connecting the p (or n) backgates to the most negative (or positive) potential in the circuit.

The intrinsic diodes at the comparator input associated with the autozeroing switch are the source of the input slew limit problem. This summing node, which is biased to a voltage approximately midscale between the supplies, must be maintained at a high impedance to work properly. For high frequency inputs or fast slewing signals such as pulses, the parasitic diodes may become forward biased, resulting in loss of charge from the coupling capacitor.

Examine the state of a comparator near the top of the conversion range, with a reference potential near V_{REF}. Before the input sampling interval begins, the coupling capacitor potential will be $V_{REF} - V_{DD}/2$. If the full-scale V_{REF} and V_{DD} are both 5 V, the summing node is initially 2.5 V below the potential of the input. In the case where a signal near ground is immediately sampled in the next clock cycle, the summing node will attempt to follow the -5-V change in the input by slewing towards -2.5 V, maintaining the original charge on the capacitor. However, the parasitic devices on the summing node will be clamped at approximately -0.7 V when the diode formed by the n-type source/drain terminal of the switch becomes forward biased. The voltage across the

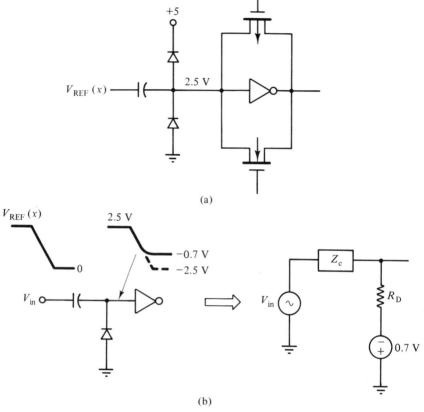

Figure 2-28   (a) CMOS comparator with parasitic diodes. (b) Equivalent circuit during clamping.

capacitor immediately changes from 2.5 V to 0.7 V, destroying the reference level.

In fact the same phenomenon occurs at a number of comparators with reference voltages that meet the following condition:

$$V_{REF}(X) - V_{in} > V_{az} + 0.7$$

For this example, all comparators with reference potentials between 3.2 and 5 V will have their reference charge removed by the forward-biased diodes. An opposite situation will occur for p-type junctions in comparators near ground that are subsequently switched to levels near V_{REF}. In many cases where input bandwidth does not exceed Nyquist limits, there will be no problem observed from this situation even though it occurs for DC and AC signals. However, some CMOS devices may be sensitive to latchup of parasitic SCRs, which can be turned on by the injection of current into the substrate.

Comparators in which diode clamping is induced do not affect the digitized result unless the signal, during a single sampling interval, returns to a voltage equivalent to the reference level of one of these stages. Otherwise, the clamped comparator is not required to detect the exact signal level, and the direction of clamping always produces the correct polarity logic output. During the clamping, however, the equivalent input of the comparator is changed as shown in Fig. 2-28b. The impedance of the input capacitor, which depends on the input frequency, is in series with the on-resistance of the forward-biased diode. Since the impedance of the diode is generally much less than the input capacitor, it is safe to assume that very little of the signal appears on the comparator input. It is also a reasonable approximation that all clamped comparators become biased to identical voltages once the diodes are turned on.

The end result is that the clamped comparators can no longer function. When the final sampled signal level is within the span of nonfunctional comparators, no active output can be produced to the thermometer encoder. Instead, the 1/0 boundary is stuck at the level of the last functioning comparator. Eventually, when the comparators return to the autozeroed mode, the low impedance of the feedback switch and comparator output are sufficient to pull the input back up and turn off the parasitic diode.

The input frequency at which the onset of clamping can be observed can be calculated when the length of the sampling period is known. For a sine input which starts at 0 V and ends at 3.2 V in the sampling period T_s, as shown in Fig. 2-29, the frequency calculation is as follows:

$$3.2 = 2.5 \cdot [1 + \sin(\omega T_s - \pi/2)]$$
$$= 2.5 \cdot [1 - \cos(\omega T_s)]$$

$$F_{clamp} = \frac{\cos^{-1}(1 - 3.2/2.5)}{2\pi T_s}$$

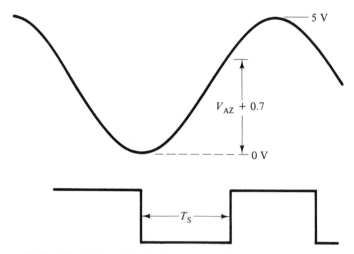

Figure 2-29   Threshold of diode clamping in CMOS autozeroed comparators.

When a 50% duty cycle clock is used, this frequency is always greater than the Nyquist rate. As an example, when T_s is 25 nsec, equivalent to a 20-MHz clock, F_{clamp} is 11.8 MHz. In some CMOS flash A/Ds which permit direct control of the sampling pulse duration, narrowing of the pulse width can be used to extend the large-signal bandwidth to higher frequencies. In CMOS A/Ds, which are capable of performing with a reduced reference voltage range (less than the sum of the autozero voltage plus diode forward voltage), the clamping problem can be totally avoided.

Bipolar A/Ds

The slewing ability of the emitter-follower buffer in the input signal path of the bipolar flash A/D limits the large-signal bandwidth of the converter. As in the case of determining small-signal bandwidth, the equivalent circuit from Fig. 2-27 shows that the buffer response is limited by the Miller capacitance of the comparator input and the output impedance of the emitter follower. For a full-scale signal swing at the input, the buffer can be analyzed as a simple RC network, similar to the settling time calculation for

CMOS A/Ds that was represented in Fig. 2-21. In this case, the following equations are applied to the comparator input:

$$V_{in} = \frac{V_{REF}}{2^{N+1}} = V_{REF} \cdot e^{-T_s/R_{eq}C_T}$$

$$T_s = R_{eq} \cdot C_T \cdot (N + 1) \cdot \ln(2)$$
$$= 0.69 \cdot (N + 1) \cdot R_{eq} \cdot C_T$$

As an example, assume the same parameters as in the earlier calculation for small-signal bandwidth, $R_{eq} = 500$ and $C_T = 20$ pF. For an 8-bit flash A/D, the settling or slewing time for a full-scale transition to $\frac{1}{2}$ LSB would be 62 nsec. This is equivalent to one-half the period of a 8-MHz full-scale sine-wave input signal.

The earlier equation for calculating the 3-dB bandwidth can be combined with the calculation above to represent the full-scale bandwidth in a more useful form. Both specifications depend on the performance of the internal buffers. It is also more common for manufacturers to specify only the small-signal bandwidth. By rearranging the terms, the equation for full-scale bandwidth results:

$$f_{FS} = \frac{1}{2T_s} = \frac{\pi \cdot f_{-3\,dB}}{\ln(2) \cdot (N + 1)} = 4.53 \cdot \frac{f_{-3\,dB}}{(N + 1)}$$

With the assumptions of the model presented above, this equation more clearly shows that in bipolar flash A/Ds the full-scale bandwidth is always substantially lower than the 3-dB bandwidth. In an 8-bit A/D, to achieve $\frac{1}{2}$-LSB settling on a full-scale transition the input bandwidth must be limited to one-half of the 3-dB bandwidth. This degradation of performance can be especially significant in pulse sampling applications, where it may be advisable to use an external S/H circuit to stabilize the input signal.

Spurious Codes, Metastable States, and Encoders

Spurious codes, also known as sparkle codes for the effect they have on digitized video displays, are grossly incorrect digital output codes that can be created by any high-speed A/D. Often this phenomenon is manifested by full-scale or all-zero outputs when the signal is actually at a point well within the limits of the A/D's range.

Metastability is a phenomenon typically associated with binary digital logic systems, particularly those using flip-flops for synchronization of signals. A latch is expected to store two distinct states, representing a logic 1 and a logic 0. All flip-flops are also capable of generating a third, indeterminate logic level between the 1 and 0 levels. This can occur when setup and hold times are violated, causing a clock edge to occur while an input signal is still in transition. The indeterminate state is described as being metastable because the condition will eventually decay, not necessarily in a monotonic fashion, to one of the valid logic levels. However, the propagation of the invalid state may cause errors before it is corrected.

In flash A/Ds, the metastability will generate spurious codes in the digital output. Analog designers are probably familiar with instability problems when using high-speed comparators. This is an example of the problem in a 1-bit A/D. In a flash A/D, the latch associated with each comparator is a potential source of metastability. Settling time in a comparator will increase exponentially as the amount of overdrive gets smaller. Specifications for A/Ds are usually based on expectations of no greater than $\frac{1}{2}$-LSB resolution. However, quantization of continuous analog waveforms inevitably results in the need to properly discriminate inputs that are small fractions of an LSB from adjacent code boundaries. In other architectures, such as successive approximation A/Ds, this lack of additional resolution may only cause individual bit errors that are partially compensated by quantizing the remaining error from the D/A in subsequent iterations on the lower bits.

When a flash A/D running at a high clock rate samples an input close to one of the reference levels, the comparator at that level will suddenly require an extended time to settle. In CMOS A/Ds, the input creates insufficient overdrive to overcome the autozeroed level at the comparator input. Since the autozeroed level is precisely at the CMOS logic threshold, the latch may then hold a signal that is neither a 1 nor a 0 and propagate it forward to the thermometer encoder. In bipolar comparators, an insufficient differential signal to the latch may have the same result.

To see the effect of the invalid logic level, examine Fig. 2-30. In the figure, the 1/0 boundary detection in the thermometer decoder is performed by a 2-input NAND gate. Assume that a signal near the second reference level causes Q_2 (and its inverse) to be indeterminate. The first comparator properly detects that the input is higher than its reference, producing a logic 1 for Q_1. Likewise, the third comparator properly detects the input as being less than its reference, resulting in a 0 for Q_3. The loss of resolution is described by the condition

$$V_{REFI} < V_{in} < V_{REF3}$$

The metastable state, since it is at the logic threshold, can result in nearly identical voltages being generated for both Q and $\overline{Q}$. In Fig. 2-30, this causes both of the adjacent thermometer decoders to generate active output signals to the encoder. Simultaneous addressing of two N-bit codes will cause a discontinuity, or glitch, in the reconstructed waveform, as in Fig. 2-31. Size and position of the glitches will depend on the design of the encoder. In a dynamic logic binary encoder, even if the metastable state settles out, the erroneous address may cause the loss of the precharged level and eliminate a logic 1 from the output. Consider the case at midscale of the A/D. For an 8-bit device the adjacent codes are

$$Q_{127} \rightarrow \quad 01111111$$
$$Q_{128} \rightarrow \quad 10000000$$

If the midscale comparators become metastable, all zeros will be generated by the logical AND of adjacent codes and will be latched at the output. The glitches become most severe at the

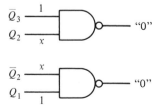

Figure 2-30 Propagation of metastable states.

Output
code

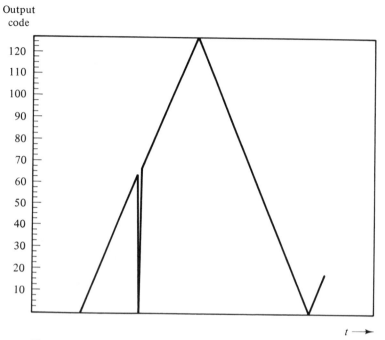

Figure 2-31   Output code glitch caused by metastable states.

MSB transition points, which are also referred to as the "major carries."

One method to suppress the generation of full-scale errors resulting from metastable conditions is the use of a Gray encoder rather than straight binary logic. Gray code provides for adjacent digital words in the encoder that differ by only one bit. Thus the simultaneous addressing of two words will result in only 1 LSB of error.

Using a Gray encoder, the example for the midscale codes of an 8-bit A/D becomes

$$Q_{127} \rightarrow 01000000$$
$$Q_{128} \rightarrow \underline{11000000}$$
$$01000000$$

The logical AND of adjacent midscale codes now results in the output of code 127 rather than code 0. This approach may tend to extend the occurrence of odd codes. The major reason that Gray encoding techniques are not used exclusively is that they require additional logic to convert back to regular binary code. Additional logic delays will slow down throughput of data, which can be especially significant in the ultrafast A/Ds, which would otherwise derive the most benefit. Table 2-1 illustrates a portion of the Gray code for an 8-bit A/D.

Much of the literature on the subject of sparkle codes in A/Ds leaves the impression that this is a random phenomenon since it is often described by a statistical model. However, this should not be interpreted as implying that there is any possibility of a comparator with high overdrive randomly generating an incorrect output when clocked within the rated conversion speed. That is why tests that are often used to predict the occurrence of sparkle codes actually push the device beyond its specified clock rate. The signal being digitized may be random, but the cause of occurrence of sparkle codes in most flash A/Ds can be directly correlated to circuit response for specific regions of input voltage matched to specific codes.

Table 2-1 Portion of 8-Bit Gray Encoding

Code	B_1	B_2	B_3	B_4	B_5	B_6	B_7	B_8
0	0	0	0	0	0	0	0	0
1	0	0	0	0	0	0	0	1
2	0	0	0	0	0	0	1	1
3	0	0	0	0	0	0	1	0
4	0	0	0	0	0	1	1	0
5	0	0	0	0	0	1	1	1
6	0	0	0	0	0	1	0	1
7	0	0	0	0	0	1	0	0
8	0	0	0	0	1	1	0	0
9	0	0	0	0	1	1	0	1
10	0	0	0	0	1	1	1	1
11	0	0	0	0	1	1	1	0
12	0	0	0	0	1	0	1	0
13	0	0	0	0	1	0	1	1
14	0	0	0	0	1	0	0	1
15	0	0	0	0	1	0	0	0

A potential cause of sparkle codes that may have some randomness associated with it is attributable to aperture errors in the sampling process. For high slew rate signals, several code boundaries may be crossed within a relatively short time interval. If the sampling clock instant (or aperture) is not exactly matched among the affected comparators, the result could be more than one active output from the thermometer decoder. As Fig. 2-32 shows, the degree of adjacent decoding that is done in the thermometer logic determines the susceptibility to timing errors. If only the 1/0 boundary is decoded, a signal crossing two adjacent

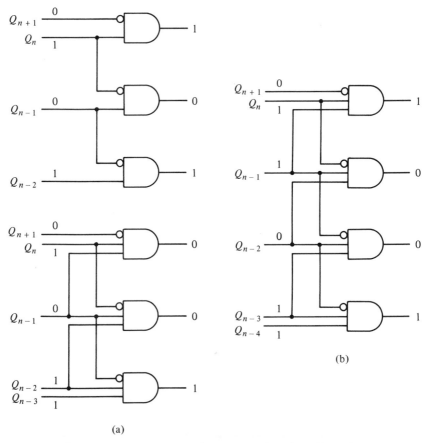

Figure 2-32   Suppression of aperture-induced sparkle codes.

code thresholds 1 LSB apart can cause ambiguous results. When three comparators are tested for the 1/1/0 condition, adjacent errors are suppressed and thresholds 2 LSBs apart must be crossed during the aperture uncertainty. The amount of timing error that is tolerable can be easily calculated for the case of a full-scale sine wave input signal:

$$V_{in} = A \cdot \sin(\omega t)$$
$$\frac{dV}{dt} = A \cdot \omega \cdot \cos(\omega t)$$

The maximum slew rate of a sine wave occurs at the zero crossing, so that dV/dt_{max} is $A \cdot \omega$ or $2\pi f A$. For 1-LSB slewing, with A equal to $2^N/2$ LSBs,

$$dt_{max} = \frac{1}{2^N \cdot \pi \cdot f_{in}}$$

Except for ultra-high-speed flash A/Ds or the case of non-Nyquist sampling or overstressed A/Ds, the generation of sparkle code errors that can be attributed to aperture uncertainty is extremely small. For an 8-bit A/D with only adjacent decoding and a 10-MHz full-scale input, dt_{max} is 124 psec. Typical timing errors would be much less than this, or other problems would become obvious from a device exhibiting such poor performance. Obviously, with a 100-MHz signal or 12 bits of resolution, this error source becomes more significant. Chapter 3 provides further discussion of slew rate limitations and aperture errors and provides a chart to find the 1-LSB threshold as a function of A/D speed and resolution.

3 ///////////////////

Modeling Error Sources: High-Speed A/D Specifications

While the performance of A/D converters has improved dramatically with the rapid advances of semiconductor technology, much confusion remains due to a lack of standardization among manufacturers in the definition of terms and test methods to describe A/D converter performance. Certainly this situation has benefited manufacturers of poorly specified or inferior products, while making the process of comparison by potential users that much more difficult. The only antidote, besides exhaustive testing, is to adhere to strict definitions that accurately describe the A/D's performance.

Some relevant standards have been issued by organizations such as the Institute of Electrical and Electronics Engineers (IEEE), the Electronics Industries Association (EIA), and the National Institute of Standards and Technology (NIST, formerly the National Bureau of Standards). For example, the JEDEC standard No. 99 published by the EIA defines some terms and symbols relating to A/D and D/A converters. There are IEEE standards, listed in the bibliography, which apply to tests for specific applications of high-speed A/D converters. Work on a general standard is also underway. Because of evolving technology and the length of time involved in producing and issuing, a standard document can quickly become out of date. Readers should be aware that manufacturer representatives volunteer to serve on the standards committees, and the final documents may sometimes reflect the particular perspective of the participants when no universal agreement exists.

While no set of parameters will satisfy every user, this chapter will describe the fundamental specifications that are relevant for high-speed A/D conversion. An attempt is also made to translate some of the most common "aliases" that a user is likely to come across when comparing manufacturer's data sheets. Often, spec-manship results in two manufacturers using different terminology to describe the same device specification. Unfortunately, it is also common for a term to have a different meaning to different manufacturers. The information in this chapter is intended to remedy this situation by preparing designers to extract the facts from the data sheets and ask the right questions when important information is lacking.

Static A/D Performance Parameters

The static performance parameters of an A/D converter describe the transfer function of the device when DC signals are present at the input. For high-speed A/Ds, this constitutes the best-case baseline performance without considering any dynamic effects. The static model is a valid starting point since the A/D conversion process itself, also referred to as quantizing or digitizing, introduces certain errors which cannot be eliminated.

Figure 3-1 illustrates the static transfer function of an ideal 3-bit A/D converter. The input range of this ideal A/D is unipolar since only positive voltages are quantized. For simplicity, the full-scale voltage is set to 8 V. By sweeping the input from zero to full scale, 2^3, or 8, distinct quantized levels are encoded in a straight binary fashion, as represented by the staircase in Fig. 3-1. The width of each step between code 001 and code 111 is exactly 1 V, which defines the resolution of this A/D and also the proportional weight of the LSB. Each code increases incrementally by $\frac{1}{8}$ of full scale. As the input voltage is swept, the risers of the transfer function, which define the transition voltages, are assumed to be of zero width, implying an instantaneous transition between con-secutive codes.

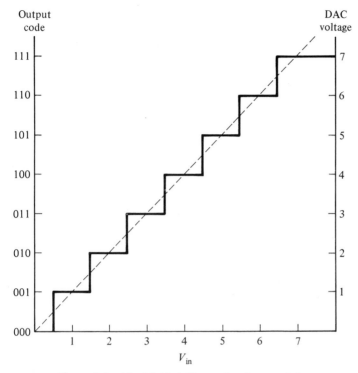

Figure 3-1 Ideal 3-bit A/D transfer characteristic.

It may be observed that this ideal transfer function is offset on the x-axis. The first code transition from 000 to 001 occurs at an input level of $\frac{1}{2}$ V, equivalent to the weight of $\frac{1}{2}$ LSB. Also, the last code transition from 110 to 111 occurs with an input of $6\frac{1}{2}$ V, which is equivalent to $1\frac{1}{2}$ LSBs less than full scale. Figures 3-2 and 3-3 should make the reasons for such a characteristic clear.

If the ideal 3-bit A/D were connected so that its output were fed directly to the input of an ideal 3-bit D/A with an 8-V full-scale range, as in Fig. 3-2, the output voltage levels would be represented by the right-side vertical axis in Fig. 3-1. The continuous straight line which passes through the center of each quantization level represents a perfect replication of the input, which would require infinite resolution from the converters.

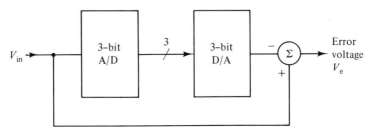

Figure 3-2   Reconstruction of quantization error.

By subtracting the D/A's quantized output from the continuous analog input, a signal will be generated which represents
the error voltage that results from the digitizing process. This
sawtooth waveform is illustrated in Fig. 3-3 exactly as it would
be observed on an oscilloscope. The sawtooth amplitude of
$\pm \frac{1}{2}$ LSB represents the quantization error of an ideal A/D. This
is the minimum error produced by an A/D of any resolution. The
reason for offsetting the transfer characteristic by $\frac{1}{2}$ LSB should
now become apparent. By employing this technique, the peak
error within the coding range of the A/D, from the first to the last
code transition, is minimized. If the first code transition occurred
at the point where the input reached a value equivalent to 1 LSB,
which is 1 V in our example, then the quantization error band
would have limits of 0 to +1 LSB. The beneficial effect of the
offset, for an ideal A/D, is to set the overall accuracy of a converter with N bits of resolution to 1 part in 2^{N+1}.

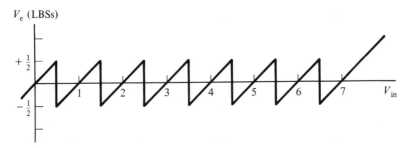

Figure 3-3   Ideal A/D quantization error.

Differential Linearity Error

The intention of the differential linearity error specification (DLE) is to describe the deviations from the ideal spacing of transition voltages in an A/D's transfer characteristic. Some manufacturers will use the term *differential nonlinearity* or just *differential linearity*. At least one flash A/D manufacturer deviates from standard data converter nomenclature by listing a specification for "code width." The code width, or code size, is synonymous for the span of input voltage which results in a given quantization level. Figure 3-1 showed that the transition voltages between consecutive codes should occur at regular intervals of $V_{fs}/2^N$, where V_{fs} is the full-scale voltage of the A/D with N bits of resolution. The uniform height of each sawtooth in the error plot of Fig. 3-3 also demonstrates this. Figure 3-4 shows the transfer characteristic of a 3-bit A/D with differential linearity errors. By examining the step widths and the sawtooth heights of the error plot, we see evidence of long and short quantization levels. To measure the differential linearity error the following equation is used (T_n represents the transition voltage of the nth quantization level).

$$\text{DLE} = \frac{T_{n+1} - T_n}{V_{fs}/2^N} - 1 \text{ LSB} \qquad (3\text{-}1)$$

In the case of the ideal A/D, all the transition voltages occur at intervals equivalent to 1 LSB, resulting in DLE of zero. In Fig. 3-4 the code 1 transition occurs at $\frac{1}{2}$ V, while the code 2 transition occurs at 1 V. In this case,

$$\text{DLE}_1 = \frac{1 - \frac{1}{2}}{8/8} - 1 = -\frac{1}{2} \text{ LSB}$$

This shows that the code 1 quantization level is short by $\frac{1}{2}$ LSB. Since the code 3 transition occurs at 2.5 V,

$$\text{DLE}_2 = 2.5 - 1 - 1 = +\frac{1}{2} \text{ LSB}$$

Code 2 results in a quantization level that is long by $\frac{1}{2}$ LSB. If a code does not appear at all in the transfer characteristic, causing a skip to the next code, then $T_n = T_{n+1}$, resulting in a DLE of

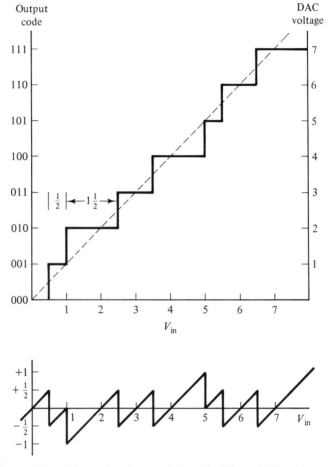

Figure 3-4   A/D transfer characteristic with differential linearity error.

-1 LSB. This would be the worst-case negative DLE of any A/D that maintains monotonic behavior. If an A/D became nonmonotonic, $T_{n+1} < T_n$, then a DLE more negative than -1 would be measured.

Although differential linearity error should appear in some form on any high-speed A/D data sheet, the prospective user cannot assume that the limits specified are measured according to Eq. (3-1). In flash A/Ds, for example, the MP7684 from Micro

Power Systems specifies differential linearity limits as a "% nominal" without clarification. TRW has published in an application note (see bibliography) a measurement technique based on the center of quantization levels rather than the transition voltages. Figure 3-5, which illustrates a portion of a possible transfer characteristic, demonstrates the ramifications of this alternative specification. The more standard definition of Eq. (3-1) would result in a DLE at codes 2 and 3 of:

$$DLE_2 = 2\tfrac{1}{3} - 2 - 1 = -\tfrac{2}{3} \, LSB$$
$$DLE_3 = 4 - 2\tfrac{1}{3} - 1 = +\tfrac{2}{3} \, LSB$$

Using a code center technique, the results would be

$$DLE_2 = 3\tfrac{1}{6} - 2\tfrac{1}{6} - 1 = 0$$
$$DLE_3 = 4\tfrac{1}{6} - 3\tfrac{1}{6} - 1 = 0$$

The use of a code-centered technique can result in the erroneous indication of zero error for quantization levels that are obviously longer or shorter than the ideal. This is particularly evident in the case of alternating long and short codes, which commonly occur in real flash A/Ds due to such asymmetrical phenomena as

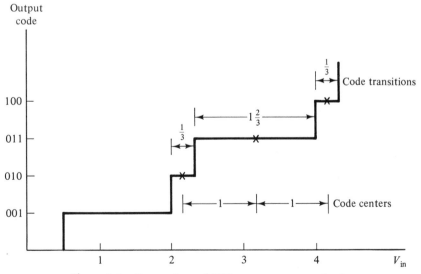

Figure 3-5 Comparison of DLE measurement methods.

the alternating switching noise induced by odd and even codes. As will be shown later in this chapter, the actual code width based on transition levels directly affects dynamic performance, such as signal-to-noise ratio.

Some manufacturers list linearity errors as a percentage of the full-scale range rather than a fraction of an LSB. To aid in translating this form of specification, refer to Table 3-1.

Offset and Gain Errors

Offset error refers to the amount by which the first code transition, from 0 to 1, deviates from the ideal position at an input equivalent to $\frac{1}{2}$ LSB. In flash A/Ds, this specification is sometimes referred to as *zero-code error, offset voltage, zero-code offset,* or *offset error bottom.* An alternate definition of this error is the amount by which the straight line through the centers of the end point quantization levels (as in Fig. 3-2) fails to pass through zero. The two definitions are identical for the ideal case where the straight line has a slope of 1, but the actual transition point can be easily measured directly. Offset error is usually specified in volts or as a fraction of an LSB and sometimes is listed as a percentage of full scale.

Table 3-1   LSB Weights as a Percentage of Full Scale

Bits	Weight of LSB (%)	Typical rounded (%)
4	6.25	6.25
5	3.125	3.1
6	1.5625	1.6
7	0.78125	0.8
8	0.390625	0.4
9	0.1953125	0.2
10	0.09765625	0.1
11	0.048828125	0.05
12	0.0244140625	0.02
13	0.01220703125	0.01
14	0.006103515625	0.006
15	0.0030517578125	0.003
16	0.00152587890625	0.002

Gain error refers to the amount by which the slope of the straight line through the transfer characteristic deviates from 1. In flash A/Ds, alternative descriptions are *scale-factor error* or a related specification of *offset error top*. Offset error top refers to the amount by which the last code transition deviates from the ideal point of full scale $-1\frac{1}{2}$ LSB. If the latter specification is provided, the slope can be derived in combination with offset error:

$$\text{Gain error} = (2^N - 2)/(T_{N-1} - T_1)$$

Gain error is usually specified as a percentage of full scale. To derive the actual transition voltages of the first and last codes, designers must keep in mind that offset and gain errors must be added to the ideal values for these points. Also, as with DLE, it is important to know whether offset voltages are measured at the code transition points or at the center of the quantization levels.

Integral Linearity Error

The purpose of the integral linearity error specification (ILE) is to describe the overall shape of an A/D's transfer characteristic. This fundamental specification is described by a variety of names and is subject to almost as many definitions. Some common aliases are *static linearity, nonlinearity, integral nonlinearity, integral linearity,* and *absolute linearity*.

The transition points of an ideal A/D, represented by the dashed lines in Fig. 3-1, should lie on a straight line. Any deviation from the straight line causes an additional inaccuracy beyond the amount due to quantization error. From these simple statements the strict definition of integral linearity error is as follows:

> *Integral linearity error is the measure of the maximum deviation of the actual transition points in an A/D's transfer characteristic from the straight line drawn between the end points (first and last code transitions). For purposes of this measurement, offset and gain errors are set to zero since they can usually be nulled out in actual use.*

This "strict definition" is the only one that can be used to give a true picture of the shape of the transfer characteristic. To facilitate a comparison of manufacturers' test methods, the more common variations and their associated pitfalls will also be described.

Figure 3-6, which once again demonstrates a case of alternating long and short codes, will be used to illustrate the ILE specification. The solid staircase depicts the actual A/D characteristic, with the numbered transition points highlighted by dots for each quantization level. These dots are also represented by the peaks of the error plot sawtooth waveform. Either set of peaks could be used. A dashed line is drawn from transition point 1 through transition point 7. The dashed-line risers indicate the transition points that would occur in an ideal A/D. Rather than intersecting the straight line, transition point 2 is offset by the equivalent of $\frac{1}{2}$ LSB in the negative direction. ILE measured at this point is $-\frac{1}{2}$ LSB, which can also be seen as the deviation in the error plot peaks between points 1 and 2. The peak-to-peak span of the error plot, which is ideally 1 LSB, is now $1\frac{1}{2}$ LSBs due to the additional error source. Similarly, the ILE at point 3 returns to zero, while at point 4 it is also $-\frac{1}{2}$ LSB. Knowing this information for each transition point, it is a simple matter to reconstruct all the steps of the transfer characteristic. For example, the error at transition point 2 tells us the width of quantization level 1, while the zero error at point 3 defines the extent of level 2. The overall specification for this 3-bit A/D is $+0$, $-\frac{1}{2}$ LSBs. As an additional exercise it may be seen that the ILE at any transition point is simply the sum of the DLE errors of all the preceding codes.

The solid diagonal line in Fig. 3-6 is the basis of one of the more popular alternative ILE definitions. This method states that ILE is a measure of the difference between the actual and the ideal quantization levels in the transfer characteristic. As before, a straight line is drawn through two end points, which are now represented by the midpoints of the first and last quantization step. In this case the distance from the midpoint of each level to the straight line is defined as the ILE.

The shortcomings of the code-centered approach should be obvious from Fig. 3-6. All of the midpoints in the example fall exactly on the line, resulting in a measurement of zero error. If

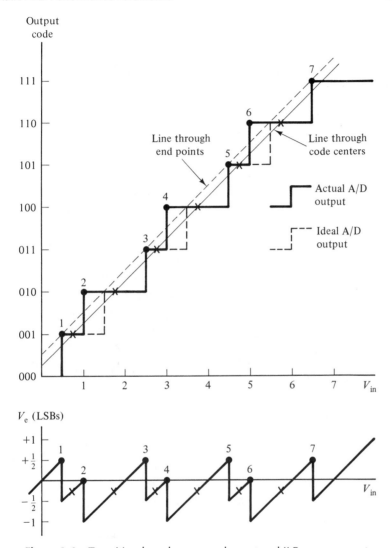

Figure 3-6 Transition-based versus code-centered ILE measurement.

the conditions of such a measurement were not known, this result might be misconstrued as indicating no error beyond the $\pm\frac{1}{2}$-LSB quantization error. Such is not the case, however, since the quantization levels are not of identical 1-LSB widths. It should be apparent that this method of describing ILE does not provide sufficient information to reconstruct the A/D transfer characteristic or to determine accuracy.

The most unclear method for measuring ILE that is often used but seldom specified clearly is the "best-fit" method. Rather than using a straight line through the end points of the transfer characteristic, this method uses a straight line that is "fit" to the actual transition points or code centers to minimize the overall error. The position of this line can be different for each sample of an A/D. A graphical example of the consequences of this approach, combined with the use of code centers, is depicted in Fig. 3-7.

For the bowed characteristic in the example, a transition-based end-point measurement from the peaks of the error voltage sawtooth results in ILE of $+0$, $-1\frac{1}{2}$ LSBs. This is evident since the error waveform has a negative peak that extends $1\frac{1}{2}$ LSBs below the ideal error band. An end-point measurement based on the span of code centers yields an ILE of $+0/-1$ LSB. A typical best-fit line to the code centers of the error waveform would balance out the ILE by drawing a line through the center of the code span, resulting in ILE of $\pm\frac{1}{2}$ LSB. Similarly, $\pm\frac{3}{4}$ LSB would be derived from the best-fit line to the transition points. The most obvious purpose of this method is to improve the appearance of the A/D error in data sheets. To have any relevance the user must be able to calibrate to the actual line used for the measurement.

Probably the most popular way to derive the best-fit line in real situations is by the "least-squares" method. This method assumes that the data being fit is essentially a linear function and seeks to develop the formula which minimizes the squared error between the real data and a straight line. The following equation describes the function which is minimized:

$$f(a,b) = \sum_{i=1}^{n}[Q_i - (aV_i + b)]^2 \tag{3-2}$$

Q_i represents the output code which results from an input voltage of V_i. The slope of the best-fit line is given by a, and the point at

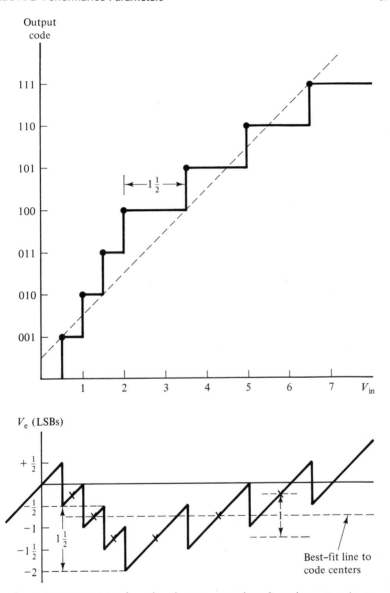

Figure 3-7 Transition-based end point versus best-fit code-centered ILE.

which the line intercepts the y-axis is given by b. Without going through all the mathematical details, if a set of input and output values are collected, a and b can be determined from the following equations:

$$a = [\Sigma\ Q_i(V_i - \overline{V})]/[\Sigma(V_i - \overline{V})^2] \qquad (3\text{-}3)$$

$$b = \overline{Q} - a\overline{V} \qquad (3\text{-}4)$$

$$\overline{Q} = 1/n\ \Sigma\ Q_i, \qquad \overline{V} = 1/n\ \Sigma\ V_i$$

The constants $\overline{Q}$ and $\overline{V}$ are the mean values of the input and output.

It should not be assumed from this simple example that the shape of the transfer characteristic can be extracted from a best-fit ILE specification. It is doubtful that a user would ever prefer this method, since it creates more complexity and tends to hide the details of a device's performance. If accuracy in a high-speed A/D conversion is important, the error budget must be calculated from the end-point transitions. For dynamic performance the actual shape of the transfer characteristic directly affects parameters such as harmonic distortion.

Power Supply Rejection Ratio

Analog designers should be familiar with the power supply rejection ratio (PSRR) specification as it is typically applied to operational amplifiers and linear circuits. In such cases PSRR is actually an AC specification which measures, as a function of frequency, the degree to which signals superimposed on the power supplies are attenuated at the output of a circuit. This measure of performance may be equally important for the A/D converter in a signal processing system, but, although PSRR is often listed on data sheet specifications, it is usually not measured in the same manner.

Typically, PSRR of an A/D converter is used as a measure of the change in the A/D's output code for a DC variation in the supply voltages. With a DC voltage which is used as the input signal, the output code should ideally remain constant with such low frequency drift in the supplies. In actuality, bias current sensitivity of an A/D's comparators may cause a change in offset voltages, which will affect the linearity, offset, and gain of the converter, resulting in a shift of the output code. Changes in the A/D's DC bias currents will also affect the dynamic response of its comparators, but this is rarely specified. The usual units of measure for PSRR are as a percentage of full scale per volt, parts per million per volt, or in decibels. The user usually has to work backwards to calculate the number of LSBs of variation. As an example, an 8-bit A/D might be specified with PSRR of 2%/V, or −34 dBs:

$$\text{PSRR}_{\%} \text{ error} = 0.02 \cdot 2^8 = 5 \text{ codes}$$

or,

$$\text{PSRR}_{\text{dB}} = 20 \cdot \log(0.02) = -34 \text{ dBs}$$

$$\text{PSRR}_{\text{dB}} \text{ error} = \frac{2^8}{10^{\frac{34\ \text{dB}}{20}}} = 5 \text{ codes}$$

This is a significant error, since 1 V peak-to-peak is equivalent to a ± 10% variation in a 5-V supply.

Dynamic A/D Performance Parameters

Signal-to-Noise Ratio

In the process of quantizing a dynamic signal, the error voltage waveform that is generated, as in Fig. 3-3, represents a noise source which corrupts the digital representation of the input signal. The RMS (root-mean squared) amplitude of this error can be easily derived. First, the size of each quantization level, equal to $V_{FS}/2^N$, will be designated as having an amplitude of q. In the ideal A/D all quantization levels are equal and spaced exactly at 1-LSB intervals. If the error signal is generated by a ramp input signal, a uniform distribution of codes results in the sawtooth with a periodicity which is designated as T. The error signal is then described by

$$v_e(t) = \frac{qt}{T} \bigg|_{-T/2}^{T/2}$$

The RMS value of this function can be calculated with the following standard equations:

$$\overline{v_e^2} = 1/T \int_{-T/2}^{T/2} (qt/T)^2 \, dt$$

$$= \frac{q^2}{T^3} \frac{t^3}{3} \bigg|_{-T/2}^{T/2} = \frac{q^2}{3T^3} \left[\frac{T^3}{8} + \frac{T^3}{8} \right]$$

$$\overline{v_e^2} = \frac{q^2}{12}$$

$$v_{e,\,RMS} = \frac{q}{\sqrt{12}} = \frac{q}{2\sqrt{3}} \qquad (3\text{-}5)$$

If a sine wave with a peak-to-peak amplitude equal to the A/D full-scale voltage is used as an input signal, its RMS voltage would be

$$V_{in,\,RMS} = \frac{2^N q}{2\sqrt{2}}$$

The RMS-to-RMS signal-to-noise ratio (SNR) for the ideal A/D is then given by

$$SNR = 20 \log \frac{2^N q/2\sqrt{2}}{q/\sqrt{12}}$$

$$= 20 \left[N \log(2) + \log(\sqrt{12}) - \log(2\sqrt{2}) \right]$$

$$= 6.02N + 10.79 - 9.03$$

$$SNR \,|_{RMS} = 6.02N + 1.76$$

This the well-known equation which relates ideal SNR to the A/D resolution. It should be intuitive that an increase in resolution reduces the error amplitude by a factor of two per bit, resulting in an increase in SNR of 6 dB per bit.

To develop a model for the RMS quantization noise in a real A/D, the example with alternating long and short codes depicted in Fig. 3-6 is used. For such a characteristic, the noise introduced by the long and short codes is at first considered separately. The error voltage from a quantization level with DLE of $+\frac{1}{2}$ LSB is described as

$$v_{\text{long}}(t) = \left. \frac{qt}{T} \right|_{-3T/4}^{3T/4}$$

The ramp extends over a quantization period equivalent to $1\frac{1}{2}$ LSBs. Similarly, for the short codes the error extends over a period of $\frac{1}{2}$ LSB:

$$v_{\text{short}}(t) = \left. \frac{qt}{T} \right|_{-1T/4}^{1T/4}$$

The squared error for the long codes is

$$\overline{v_{\text{el}}^2} = \frac{2}{3T} \int_{-3T/4}^{3T/4} (qt/T)^2 \, dt$$

$$= \left. \frac{2q^2}{3T^3} \frac{t^3}{3} \right|_{-3T/4}^{3T/4}$$

$$\overline{v_{\text{el}}^2} = \frac{3q^2}{16}$$

$$v_{\text{el, RMS}} = \frac{\sqrt{3}q}{4} = \frac{3}{4} \frac{q}{\sqrt{3}}$$

The final result shows that the RMS noise contributed by a code that is 50% longer than the ideal is 50% larger than the amount shown in Eq. (3-5). Similar equations for the short codes are as follows:

$$\overline{v_{\text{es}}^2} = \frac{2}{T} \int_{-T/4}^{T/4} (qt/T)^2 \, dt$$

$$= \left. \frac{2q^2}{T^3} \frac{t^3}{3} \right|_{-T/4}^{T/4}$$

$$\overline{v_{\text{es}}^2} = \frac{q^2}{48}$$

$$v_{\text{es, RMS}} = \frac{q}{4\sqrt{3}}$$

The noise generated by a code that is 50% short is half the noise that results from an ideal quantization level. Over the full-scale range of the A/D, it will be assumed that there are an equal number of long and short codes. That repetitive pattern that exists over the space of 2 LSBs is depicted in Fig. 3-8. The noise from the long and short codes does not average to zero since, to arrive at a total for the squared error, the two sources of noise must be appropriately weighted by their probability of occurrence. From Fig. 3-8 it can be seen that long codes will occupy three-fourths of the range determined by $2^N - 2$ quantization levels, with the short codes occupying only one-fourth of the total.

The alternating code model leads to the following result:

$$\overline{v_e^2} = \frac{3}{4}\overline{v_{el}^2} + \frac{1}{4}\overline{v_{es}^2}$$

$$= \frac{3}{4} \cdot \left[\frac{3}{2} v_e\right]^2 + \frac{1}{4} \cdot \left[\frac{1}{2} v_e\right]^2$$

$$v_{e,\,RMS} = 1.32\ q/\sqrt{12}$$

The overall effect of alternating codes is a 32% increase in quantization noise. The degradation in SNR is then

$$SNR|_{RMS} = 20 \log \frac{2^N q/2\sqrt{2}}{1.32 q/\sqrt{12}}$$

$$= SNR_{ideal} - 20 \log (1.32)$$

$$SNR|_{RMS} = 6.02N - 0.65\ dBs$$

The increased quantization error reduces SNR by 2.41 dBs. A general equation for the alternating code model with variable DLE error is

$$v_{e,\,RMS} = q/\sqrt{12} \cdot \sqrt{\left|\frac{(1 + DLE)^3 + (DLE)^3}{2}\right|}$$

This equation is used in Fig. 3-9 to illustrate the degradation in SNR versus DLE error. This should serve as a guideline to relate the dynamic performance that can be expected in an A/D with a given static performance characteristic, since the long and short codes do tend to balance out. In an actual device, a more accurate RMS summation of errors would be complicated by the real distribution of code sizes.

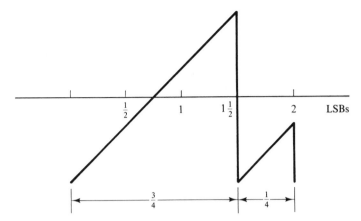

Figure 3-8   Percentage of quantization range for long and short codes.

The usual method of measuring SNR involves an analysis with a fast Fourier transform (FFT) algorithm, which will be discussed in Chapter 6. The FFT analysis of sampled data yields the same information as would be obtained by using an analog spectrum analyzer on a continuous waveform (e.g., noise and harmonic distortion). One of the points to be wary of when reviewing SNR specifications is that some manufacturers separate these dynamic errors, reporting SNR and THD (total harmonic distortion) separately. Others report the ratios of both signal-to-noise and signal to noise + distortion. Removing the error signals that are harmonically related to the input will artificially inflate the SNR. Occasionally, a manufacturer will measure the peak signal to RMS noise. Be aware that in such cases 3 dBs must be added to the ideal SNR.

As part of the process of separating harmonic distortion from noise, spectral averaging is sometimes employed. By accumulating multiple sets of data as input for an FFT, random signals will tend to be averaged out. This allows small amplitude harmonics to be detected while lowering the noise floor in the data. Be aware that if the result of this process is used to report SNR for the A/D, it will give a higher value than the user can expect to achieve in any single measurement.

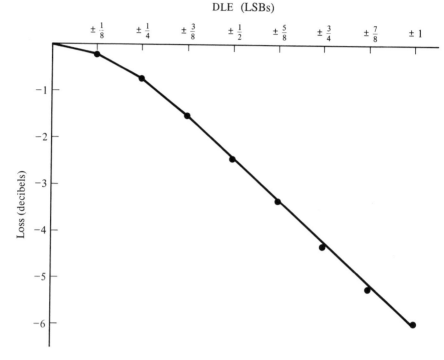

Figure 3-9   Loss of SNR versus DLE, alternating code model.

Effective Bits

The measured SNR and the ideal model for quantization noise are sometimes used to calculate a figure of merit known as effective bits. Two popular calculation methods are described by the following equations:

$$\text{Effective bits} = \frac{\text{measured SNR} - 1.76}{6.02}$$

or

$$\text{Effective bits} = N - \log_2 \frac{\text{measured RMS noise}}{\text{ideal noise}}$$

The SNR model of an ideal A/D, which leads to the first equation, is based on a uniform distribution of codes. Since this situation does not typically exist in reality, the second equation is sometimes used. The "ideal noise" in this case is the amount that results when an ideal sine wave is best-fit to the measured data. An algorithm to minimize the mean-squared error is often used, similar in concept to the ILE measurement discussed earlier. For sine waves, however, the parameters of amplitude, frequency, phase, and DC offset must all be estimated. (A method of performing a best-fit sine wave calculation is described in Chapter 6.)

This figure of merit is most useful when it is measured over the full bandwidth of the A/D. Over a range of frequency it can be used to illustrate the degradation in performance due to dynamic effects.

Aperture Errors

Since all flash A/Ds provide an intrinsic sampling function for dynamic input signals, the aperture errors which would otherwise be determined from the S/H (sample and hold) specifications are important. There are three such parameters which are usually listed: *aperture delay, aperture uncertainty,* and *aperture jitter.*

Aperture delay, while not really an error source, refers to the time between an input clock edge that defines the sampling period and the actual switching instant internal to the A/D. Depending on the A/D's architecture, this timing could define the beginning or end of the time when the input signal is actually digitized. Conceptually, at least, once this time is known, events external to the A/D can be synchronized through use of the appropriate delays.

Aperture uncertainty and aperture jitter, while not necessarily synonymous, are often grouped together as a single error source in data sheets and sometimes referred to simply as aperture error. It is practically impossible to separate the two effects, one due to random variations and the other dependent on the signal itself. In flash A/Ds the mismatch between internal circuits, both analog and digital, may result in a variation of the sampling instant or aperture across the range of comparators. This would properly be defined as aperture uncertainty and would be directly related to the signal's level as well as to its slew rate. Aperture jitter refers to the random variations in the sampling point due to noise sources in the A/D's circuitry. Thermal noise, for example, when added to the input of a clock driver would cause random variations at the logic threshold resulting in an output transition with variable delay relative to the input.

The direct measurement of this error source, which is described in more detail in Chapter 6, is very difficult, and a specification is usually arrived at through a process of estimation. To illustrate the effect of this error source, Fig. 3-10 shows how sampling uncertainty and jitter result in a lack of accuracy in determining the voltage level of a high slew rate signal.

One of the difficulties in the measurement of this parameter is that for it to be directly observable, repetitive sampling of a fixed point must result in more than one code at the output. Errors less than 1 bit require that the input signal voltage at the sampling instant and the DC transition points of the corresponding quantization level be known precisely in order to attribute any additional quantization error to aperture effects. Recall that the ideal staircase A/D transfer function makes instantaneous monotonic transitions from one code to the next. In reality there is a finite uncertainty in the transition voltage, even for static input signals, due to limited resolution and settling time in the A/D's comparators as well as noise effects. This error, which could be

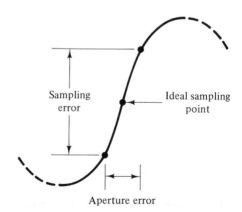

Figure 3-10 Sampled error due to aperture errors.

defined as precision, is never specified on data sheets, but it makes it nearly impossible to make direct dynamic measurements of signals near code transition points.

A quality high-speed A/D should be expected to be capable of digitizing high slew rate signals at its input. A full-scale sine wave can be used to represent a worst-case condition, and its slew rate can be used to accurately determine the amount of aperture error which is tolerable:

$$\text{Slew rate} = \frac{dV}{dt} = \frac{d}{dt} [V_{fs}/2 \cdot \sin (\omega t)]$$
$$= \pi \cdot f \cdot V_{fs} \cdot \cos (\omega t)$$

The maximum slew rate occurs at the ''zero crossing'' (actually midscale for a unipolar A/D) where $\cos (\omega t)$ is 1, so that

$$\frac{dV}{dt_{MAX}} = \pi \cdot f \cdot V_{fs}$$

As a guideline to how much aperture error is tolerable, assume that dV is equivalent to $\frac{1}{2}$ LSB or $V_{fs}/2^{N+1}$:

$$\text{Aperture error} = dt = \frac{V_{fs}/2^{N+1}}{\pi \cdot f \cdot V_{fs}}$$
$$dt_{MAX} = \frac{1}{\pi \cdot f \cdot 2^{N+1}}$$

This equation relates the amount of aperture error to the input frequency and required resolution. The following equation is an example for an 8-bit flash A/D with a 10-MHz input:

$$dt_{MAX} = \frac{1}{512 \cdot \pi \cdot 10^7}$$
$$dt_{MAX} = 62 \text{ psec}$$

It is not uncommon to see aperture error on the order of 20 psec specified for such an A/D.

Figure 3-11 provides a log–log chart which can be used to determine the aperture error that would result in $\frac{1}{2}$ LSB of sampling error at various input frequencies for A/Ds of 6 to 12 bits of resolution.

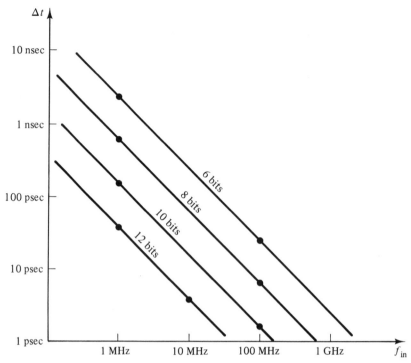

Figure 3-11   One-half-LSB aperture error limits versus f_{in}.

Input Bandwidth

The input bandwidth specification should indicate the effective attenuation and phase shift of high-frequency signals that occur due to limitations in the internal signal paths of high-speed A/Ds. This parameter is often described as *3-dB bandwidth*, while related specifications are *full-power bandwidth, small-signal bandwidth*, and *large-signal bandwidth*. In some cases the 3-dB point will occur well past the point where the total A/D is no longer functioning properly. Full-power bandwidth, as typically defined, gives an indication of the point at which gross malfunction occurs, such as missing or spurious codes. This parameter by itself gives no information about dynamic performance prior to the onset of failure.

For a worst-case analysis, bandwidth must be specified for full-scale input signals. When small-signal bandwidth is specified, conditions of signal amplitude should be given or a separate parameter provided so that slew rate limitations can be determined. Other alternatives that have been suggested for bandwidth specifications include the 3-dB point of SNR. This also gives only a partial measure of performance since large dynamic errors can be hidden in the RMS noise.

The input bandwidth should be measured separately from any buffer amplifier which may drive the A/D. Input amplitude should be held constant while the peaks of the digitized output are evaluated. This can be done directly using computerized techniques or by examining the reconstructed output of a high-speed D/A. As in any analog signal processing circuit, bandwidth limitations of the A/D affect both the amplitude through attenuation and timing accuracy through phase shift. When the A/D's frequency response is governed by a single dominant pole, the following equations can be used:

$$A(f) = \frac{1}{\sqrt{[1 + (f/f_3)^2]}}$$
$$\theta(f) = \tan^{-1}(f/f_3)$$
$$t_e(f) = \frac{\theta(f)}{360°} \cdot \frac{1}{f}$$

Where attenuation factor is $A(f)$, phase shift is $\theta(f)$, and time shift error for a sine wave input is $t_e(f)$. The 3-dB frequency is given as f_3.

Table 3-2 demonstrates the relationship between amplitude and phase errors and the 3-dB frequency. As an example, to maintain a peak accuracy of 8 bits (i.e., $A(f) < 0.4\%$) the table shows that the 3-dB bandwidth must be more than ten times the signal frequency. Such a wide dynamic range is often difficult and expensive to achieve.

In some applications the timing accuracy of uniformly sampled high-frequency spectra is important. In such cases, and also for pulse amplitude measurements, the delay introduced by phase

Table 3-2   Amplitude and Phase Error
versus Frequency

f/f_3	$A(f)$ %	$A(f)$ dBs	$\theta(f)$
.05	−0.12	−0.01	2.9°
.1	−0.5	−0.44	5.7°
.2	−1.9	−0.17	11.3°
.3	−4.2	−0.37	16.7°
.4	−7.2	−0.65	21.8°
.5	−10.6	−0.97	26.6°
.6	−14.3	−1.34	30.9°
.7	−18.1	−1.73	35.0°
.8	−21.9	−2.15	38.7°
.9	−25.7	−2.58	42.0°
1.0	−29.3	−3.00	45.0°

shift can be significant. The overall effect on a sine wave due to input bandwidth limitations can be summarized as

$$V(t)_{\text{eff}} = \frac{A}{\sqrt{[1 + (f/f_3)^2]}} \sin [\omega t + \tan^{-1}(f/f_3)]$$

Differential Gain and Differential Phase Errors

These two specifications are specifically associated with the quantization of video signals. The textbook definition for differential gain (DG) is "the percentage difference in the output *amplitude* of a small high-frequency sine wave at two stated levels of a low-frequency signal upon which it is superimposed." Differential phase (DP) follows a similar definition where the change in the output phase at two points is measured in degrees. For the composite video format used in North America, standards and test signals are defined by the NTSC (National Television Standards Committee). A typical test signal is illustrated in Fig. 3-12.

The composite video signal, in Fig. 3-12 contains all the picture and control information needed to display each line of each frame on a television monitor. The example test signal is equivalent to one complete horizontal scan line. The pulse portion represents the horizontal sync, followed by a burst of the color subcarrier at 3.58 MHz. Color information in the picture is transmitted by

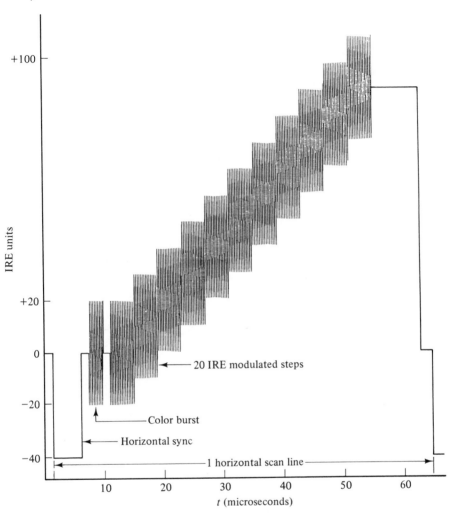

Figure 3-12   NTSC standard video test signal.

amplitude and phase modulation of this subcarrier frequency, which adjusts the intensity and hue. The composite color signal is referred to as chrominance.

The brightness of the video image is controlled by varying the DC level of the modulated waveform, and it is sufficient to create a black-and-white picture. The monochrome signal is also referred to as luminance. Measurements of DG and DP are important to indicate the degree of undesirable interaction which may occur between the luminance and chrominance. When applied to A/Ds, the DG and DP specifications measure the dynamic linearity that is maintained while digitizing the complex video waveform.

Limits in the amplitude of the various elements of the composite video signal are measured in IRE (Institute of Radio Engineers) units. One volt is equal to 140 IRE units. To measure DG and DP, the chrominance test signal with a constant peak-to-peak amplitude of 40 IRE and a constant phase is typically superimposed on 10 DC steps. As this waveform is swept through the full-scale range of an A/D, factors such as ILE and quantization error will contribute to distortion in the amplitude of the AC signal. The extent of this amplitude variation is the differential gain error. Phase nonlinearity resulting from limited input bandwidth and aperture errors distorts the timing of the video signal, resulting in differential phase errors. A more thorough analysis of the quantization of video signals is beyond the scope of this book, but the references in the bibliography provide additional information for those who are interested.

In a 100% saturated video signal, the picture content extends from −33 to +133 IRE units, for a total of 166 IRE. For maximum dynamic range, these limits are used to set the zero and full-scale of the A/D. The error contributed by quantization noise can then be calculated from the bit weighting of $166/(2^N-1)$ IRE. As an example for an 8-bit A/D, the "underflow" at $\frac{1}{2}$ LSB would be set to −33 IRE, with "overflow" at +133 IRE. Each quantization level is then equivalent to 0.651 IRE per LSB.

The specification of differential gain that is reported in data sheets is typically the RMS quantization noise that results from

Table 3-3   Ideal DG and DP Errors

Bits	DG (%)	DP
6	7.86	4.50°
8	1.94	1.11°
10	0.48	0.28°
12	0.12	0.07°

digitizing the standardized test signal shown in Fig. 3-12. This method of measurement is similar to the traditional analog technique of using a vectorscope, an instrument that does not readily provide instantaneous values of peak error. By calibrating the A/D to this signal, the following equation for differential gain results:

$$DG = \sqrt{2}/2 \cdot \frac{\dfrac{140 \text{ IRE}}{(2^N - 1)}}{20 \text{ IRE}} \cdot 100\%$$

Similarly, for differential phase:

$$DP = \sqrt{2}/2 \cdot \frac{\dfrac{140 \text{ IRE}}{(2^N - 1)}}{20 \text{ IRE}} \cdot \frac{180}{\pi} \text{ deg.}$$

Table 3-3 summarizes the theoretical limits of RMS DG and DP errors versus A/D resolution. For the data in the table, the only factor determining error is the number of bits in the A/D. However, as the analysis in the reference by Felix points out, worst-case instantaneous errors are affected greatly by the de-

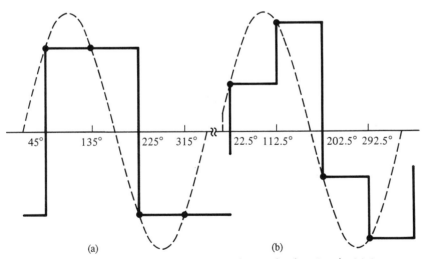

Figure 3-13 Change in gain versus phase in digitized video signals: (a) 4× oversampling, phase = 45°, and (b) 4× oversampling, phase = 22.5°.

gree of oversampling in the digitized video signal. This occurs because the peak-to-peak amplitude of the digitized samples depends on the phase relationship between the modulated signal and the sampling clock, while the quantization error is fixed. This is illustrated by the example in Fig. 3-13. Oversampling rates of at least four times the color subcarrier frequency are typically used in digital video.

When reading an A/D data sheet for the DG/DP specs, the designer needs to know what the conditions of the input signal and sampling clock are. It is not uncommon to see specifications which seemingly violate the theoretical limits shown in the table. In such a case, the manufacturer should specify if the error listed is to be interpreted as an addition to the theoretical error.

4 /////////////////

Support Circuits for
High-Speed A/Ds

The next step after selecting the flash A/D for a high-speed conversion system is to design the support circuitry that will be required to make it perform up to its specifications. Choices of peripheral components extend from opamps, buffers, and voltage references to decoupling capacitors and sockets. This chapter will provide the background information and general guidance to help in making these decisions, but the manufacturer's data sheet should always be consulted for specific recommendations that may apply to a particular A/D. Also included in this chapter are recommendations for circuits that can actually improve the performance of a device by overcoming some of the intrinsic limitations of flash A/Ds.

Reference Circuits

An assortment of devices are available for the generation of reference voltages in high-speed A/D applications. Three terminal regulators, zener diodes, and bandgap references can all be used depending on the precision and temperature drift requirements of the application. Whatever technique is employed, in most situations it is recommended that separate analog supplies be used for the reference circuit. Using the digital supplies can feed back switching noise from the A/D.

In some cases regulator components can supply sufficient current to drive a flash A/D ladder directly. In most applications, however, it is preferable to use an opamp buffer with the addition of an emitter follower to assure a low output impedance to the A/D. An example of this circuit for supplying a negative voltage to the bottom of a resistor ladder is shown in Fig. 4-1. The drive transistor is connected to the opamp in the unity gain configuration. The opamp is chosen for low noise and offset drift, while the transistor is used to reduce the load on the amplifier's output stage.

Offset and Gain Adjustments

In a flash A/D the actual end-point transitions at the first and last codes will deviate from the ideal threshold voltages of (REF$-$ $+ \frac{1}{2}$ LSB) for code 1 and (REF$+$ $- 1\frac{1}{2}$ LSB) for code 2^N-1. In devices which use bipolar transistors, these offsets result

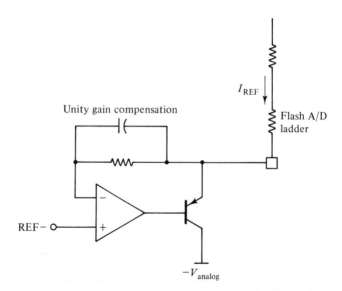

Figure 4-1   Typical low-impedance drive circuit for flash ladder.

from a combination of two factors. One factor is the offset voltage in the comparator or preamp, which directly adds to the input. The second factor, which will usually dominate, is the voltage drop which occurs in the current path through the reference ladder, flowing from the input of the first (or last) comparator to the external reference voltage source. With a total ladder resistance which typically ranges from less than 100 Ω up to 300 Ω, a voltage step of 1 LSB is generated across a segment of only 1 Ω or less. It is extremely difficult under such circumstances to provide a current path for the reference voltage from the package pin onto the integrated circuit and then to the comparator, with a series resistance equivalent to only $\frac{1}{2}$ LSB.

In some applications it may be adequate to adjust the terminal voltages applied to REF− and REF+ in order to compensate for the additional voltage drop in the parasitic series resistance. This technique can shift the internal reference tap potential to a level which results in the proper transition voltage. However, changes in operating temperature will cause offset drift because the reference resistance and internal offsets and bias currents will be affected differently. For this reason, some A/Ds include both drive and sense pins for the end points of the ladder. This allows a Kelvin connection to be made, as shown in Fig. 4-2.

In the circuit of Fig. 4-2, the sense pin is a second direct connection to the first comparator's reference input, but it is not used to supply current to the ladder. An operational amplifier is connected in the voltage follower configuration to drive the end of the ladder. By applying feedback around the parasitic resistance, from the sense pin to the inverting input of the opamp, the threshold voltage of the first comparator will be forced to follow the reference voltage V_{REF-}.

The Kelvin adjustment circuit can also be used to adjust CMOS A/Ds if drive and sense pins are provided. However, the effect will not be completely achieved, because the source of offset is different. Besides voltage drops in the parasitic resistance, CMOS A/Ds have offsets which are caused by the feedthrough of MOSFET turn-off transients which occur in the sampling process. These errors will also directly add to the input signal

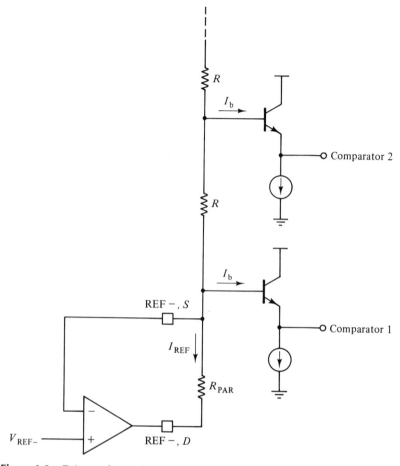

Figure 4-2   Drive and sense lines can be used as a Kelvin connection to cancel the parasitic voltage drop in the reference current path.

when they propagate to the input capacitor and summing node of the autozeroed comparator. Driving the ladder from a voltage follower will only compensate for the DC component of offset error in CMOS flash A/Ds. Unless the feedthrough transients are compensated separately, temperature drift may still occur because the parameters which determine their magnitude will change with temperature.

Linearity Trimming

Integral linearity errors in flash A/D converters are frequently manifested as a bowed or piecewise linear characteristic rather than a straight line through the end points. This may be due to cumulative resistor matching errors in the reference ladder, crossover mismatch between A/D segments, or to other non-linear phenomena such as voltage drops caused by incremental base currents flowing in bipolar devices. Fortunately for the user, external adjustments can be made to reduce this nonlinearity. Most flash A/Ds provide access to the end points of each intermediate segment of the reference ladder—at the quarter or midscale points of the A/D's range. The sources of INL error cause these points to deviate from the ideal linear transfer function. By actively driving the reference ladder from an adjustable voltage source, errors at the intermediate end points can be nulled out, thus improving the composite transfer function. The effect of the improvement is illustrated in Fig. 4-3.

One simple method of improving INL is to parallel the reference ladder of the A/D with precision resistors or a potentiometer. This scheme for a midscale reference adjustment is represented in Fig. 4-4. A limitation of the parallel resistor approach is that the compensation will not hold over changes in temperature, since the temperature coefficients of the internal and external elements will not be matched. This point is demonstrated in the following example.

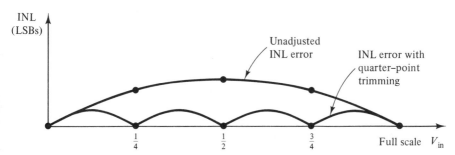

Figure 4-3 Active trimming of quarter-point reference taps reduces integral non-linearity.

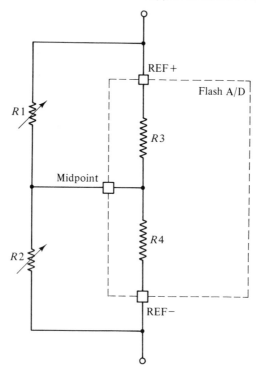

Figure 4-4 External resistors can be used to adjust integral linearity at intermediate taps on flash A/D ladder.

Assume that the internal reference resistor segments $R3$ and
$R4$ are mismatched, causing a shift in the midscale voltage. This
error will be nulled out when the external resistors are adjusted
so that the parallel combinations are equal: $R1 \parallel R3 =
R2 \parallel R4$. Adjusting for

$$\frac{1}{R_T} = \frac{1}{R1} + \frac{1}{R3} = \frac{1}{R4} + \frac{1}{R2}$$

which can be reduced to the following equation:

$$\frac{R4 - R3}{R4 \cdot R3} = \frac{R1 - R2}{R1 \cdot R2}$$

The form of this last equation is intended to emphasize the
purpose of the external resistors in establishing an equal, but

opposite, proportional difference in resistor segments. One obvious solution which achieves matching of the two halfs is to adjust the external resistors until $R1 = R4$ and $R2 = R3$. However, assume that the internal and external resistors have different temperature coefficients of resistance (TCR) such that:

$$\text{TCR of external resistor} = \alpha$$
$$\text{TCR of internal resistor} = \beta$$

$$R1(T) = R1_o(1 + \alpha T)$$
$$R2(T) = R2_o(1 + \alpha T)$$
$$R3(T) = R3_o(1 + \beta T)$$
$$R4(T) = R4_o(1 + \beta T)$$

$$\text{error} = \frac{R1(T)\|R3(T) - R2(T)\|R4(T)}{R1(T)\|R3(T) + R2(T)\|R4(T)}$$

Without going through all the algebra, the equation for the midscale drift error in this example (where $R1_o = R4_o$, $R2_o = R3_o$) can be reduced to an expression which clearly illustrates the dependence on mismatch in temperature coefficients:

$$\text{Error} = \frac{(R3_o - R4_o) \cdot (\alpha - \beta) \cdot T}{(R3_o + R4_o) \cdot (2 + \alpha T + \beta T)}$$

A plot of this function over an extended temperature range, using some typical IC parameters, is illustrated in Fig. 4-5. An error of 0.3% is approximately $\frac{3}{4}$ LSB in an 8-bit A/D. An improved adjustment circuit, shown in Fig. 4-6, negates the effect of the mismatch in temperature coefficients. In this circuit, the external trimming network is buffered from the internal resistor ladder. By using a unity gain operational amplifier to drive the flash A/D, a voltage source will be formed which maintains the adjusted resistor ratio independent of differences in temperature coefficients.

Using Track/Hold Circuits on Input Signals

Almost every flash A/D data sheet includes a statement claiming a built-in sample/hold function or a reference to the capabilities of the device to operate without an external track/hold. While

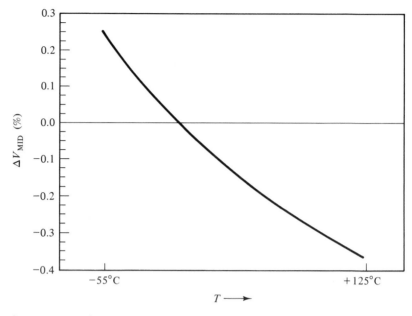

Figure 4-5 Drift in midpoint adjustment due to mismatched TCR of parallel resistors.

this claim may be legitimate subject to the performance limitations of the particular device as discussed in Chapter 2, it would be equally true to state that any flash A/D's dynamic performance can be improved by adding an external track/hold. All nonflash A/D architectures absolutely require track/hold circuits to achieve any significant input bandwidth, including two-step subranging architectures which employ flash A/Ds internally.

In fact, in many ways the track/hold function is the key to the upper limit in high-speed A/D conversion. It is through clever architectures such as pipelined and interleaved A/Ds (which depend on T/Hs to capture the signal) that the highest effective sampling rates and bandwidth are always achieved. By adding a T/H to a flash A/D system, it is possible to effectively extend the converter's DC performance to higher frequencies. Some of the factors which limit the A/D's dynamic performance will be nullified, such as small-signal input bandwidth, dynamic nonlinearity, and aperture jitter. The performance requirements for these specifications will be transferred to the T/H. Other factors, such as large-signal bandwidth and slew-rate limitations, may not necessarily benefit as much from a T/H since they still depend on the converter's ability to recover from full-scale voltage swings at the input.

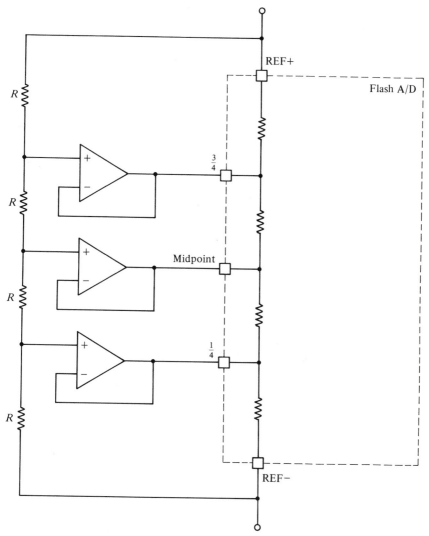

Figure 4-6 Improved method for linearity adjustment buffers trimming components from internal components.

One of the most important benefits of a T/H is in maintaining accuracy. This can be particularly important in applications where pulse height is being measured. In such situations, the gain-bandwidth and settling time of the A/D's comparators with the high spectral content of the input pulse can severely limit accuracy.

There are T/H devices available today which are designed specifically for flash A/D applications. Such devices are suitable for high-speed A/Ds with up to 10 bits of resolution. These devices are manufactured both as monolithic ICs and as multichip hybrid microcircuits, which allows for factory trimming to meet performance specifications without additional adjustments by the user. A typical architecture of such a device is represented by the MN379 from Micro Networks, which is shown in Fig. 4-7. In the tracking mode, these devices act as unity gain buffers and are capable of directly driving the capacitive load from the input of flash A/Ds. In the hold mode a fast diode bridge circuit acts as a switch, isolating the input from the output, with the sampled signal preserved on the storage capacitor.

Comparing tests of a flash A/D with and without a T/H at the input can dramatically demonstrate the performance improve-

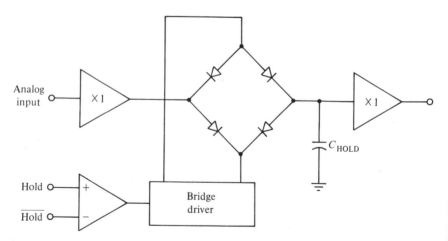

Figure 4-7 Architecture of the MN379 from Micro Networks, a fast T/H designed for flash A/D applications. (Reproduced courtesy of Micro Networks, division of Unitrode Corporation, Worcester, MA.)

ment which may be achieved. Figure 4-8 illustrates such a comparison for a bipolar 8-bit flash A/D operating at a 24-MHz sampling rate with a 10-MHz input signal. Without the T/H in the circuit, missing codes have caused large nonlinearities at the zero crossings. Because of these large errors, the SNR and harmonic distortion are also poor. With the T/H in the circuit, the zero crossings return to the performance that would be expected from the static linearity specification, SNR improves by 17 dBs, and harmonic distortion is reduced by greater than 20 dBs. While the impact of these improvements is enhanced by the fact that they were obtained on a device which is basically malfunctioning with a signal well below the Nyquist rate and without a T/H, the example gives a valid indication of the degradation in dynamic performance which could occur and how it can be remedied.

Probably the most difficult aspect of applying a T/H to a flash A/D is properly setting the timing relationship between the two devices. As is the case in many situations, there are differences in application and performance depending on whether a bipolar or CMOS flash A/D is being used. In both cases, however, the goal is to have the T/H acquire the signal during the interval that the A/D is *not* sampling its input, the strobe or latching phase for the bipolar A/D, or the autozeroing interval for the CMOS A/D.

A timing diagram which will illustrate the important relationships between the T/H and A/D clocks is shown in Fig. 4-9. The proper timing is established by considering the performance specifications that are germane to the T/H circuit. In the figure, the falling edge of the T/H clock sets it to the tracking mode. Near this time instant the rising edge of the A/D clock removes it from the sampling mode. There will be a delay t_1 between the application of the hold signal and the actual return to tracking of the T/H. Proper design will guarantee that the *aperture delay* of the A/D, t_5, has expired before the signal from the T/H starts slewing to its new level.

For T/Hs, a key specification is the *acquisition time t_2*, which is properly defined as the total time in track mode for the T/H to slew to the new input level with the desired accuracy. This parameter will determine the minimum pulse width for the autozero or strobe cycle of the A/D. Some manufacturers measure this

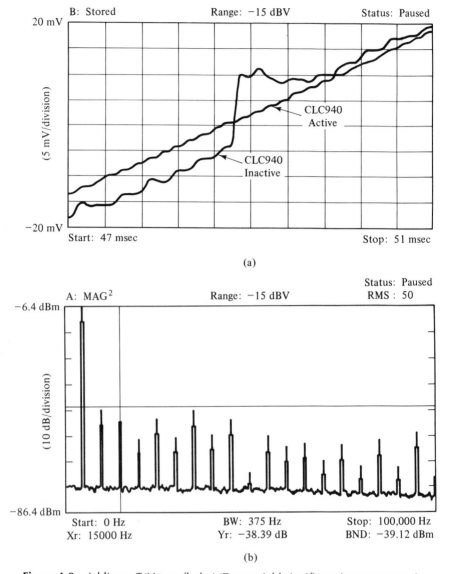

Figure 4-8 Adding a T/H to a flash A/D can yield significant improvements in dynamic performance. (a) Without the CLC940 T/H to stabilize the input signal, sampling a 10-MHz sine wave at a 24-MHz clock rate produces a 12-LSB jump at the zero crossing. (b) Harmonic distortion is −32.8 dB and SNR is 31.3 dB for the 24-MHz flash A/D without a T/H. (c) With the CLC940 T/H, dynamic performance improves to −53 dB harmonic distortion and 48.3 dB SNR. (*Figure continues.*) [Reproduced with permission from D. Potson (1987), "Track and Hold Amplifiers Improve Flash A/D Accuracy," Application Note TH-06, Comlinear Corporation, Fort Collins, CO.]

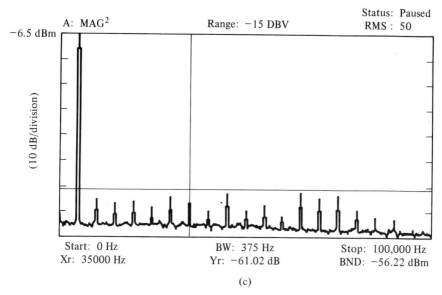

Figure 4-8 *(continued)*

parameter separately from the *hold-to-track delay time* t_1, so that the two parameters must then be added together to determine the timing requirements.

After application of the hold signal, the T/H will have its own aperture delay specification t_3, which is defined as the time required to terminate the tracking mode. Some manufacturers will separate this parameter from the *track-to-hold settling time* specification t_4. For flash A/D applications, it is not necessary that the signal be completely settled at the beginning of the A/D's sampling interval, only that the slewing be completed and the T/H is clearly in the hold mode. This is accomplished by considering the delay before sampling actually begins in the A/D t_6, which should be set to occur after the aperture delay of the T/H.

Among the other T/H parameters which must be considered is *gain,* which actually represents a loss since it is always less than 1.0 from the "unity gain" buffers. Gain of the A/D converter must be adjusted in consideration of this factor.

The *pedestal error* is the result of the feedthrough of switching transients onto the hold capacitor, which causes the T/H output

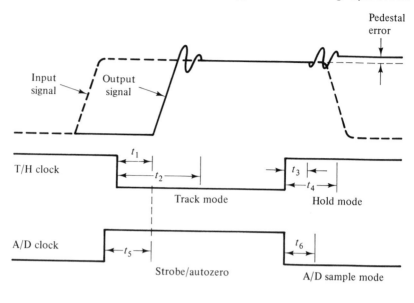

t_1 = T/H hold–track delay t_3 = T/H aperture delay
t_2 = T/H acquisition time t_4 = T/H settling time
t_5 = A/D aperture delay t_6 = A/D sampling delay

Figure 4-9 Timing diagram for application of T/H with flash A/Ds.

in the hold mode to deviate from the actual signal level at the end of the tracking interval. Besides representing an error in accuracy, the important consideration with pedestal error is how it varies with the input signal level. A T/H linearity specification must not be limited to the effects of the buffer amplifier. Linearity in the final held voltage will ultimately be determined by the combination of the switch and buffer.

Another important T/H specification is *feedthrough attenuation* or *hold-mode feedthrough*. This parameter measures the degree of isolation provided from the input to the output of the T/H in the hold mode. Any AC signals which are coupled to the A/D input must be kept to a fraction of an LSB to avoid errors in the quantization process. In subranging A/Ds, this is especially important, since the signal being digitized must not change between the coarse and fine conversions.

Applying Buffers to Drive the A/D's Input

The growth in performance of high-speed A/Ds has been accompanied by an increase in the number and type of components which are available for conditioning of the signal to be digitized. Besides T/Hs, there are wide-band, high-current, unity-gain buffer amplifiers and operational amplifiers, which are designed specifically for driving flash A/Ds. These devices are available in both hybrid and monolithic form.

A starting point in the process of selecting a buffer amplifier to drive a flash converter input is to determine the bandwidth which is required for a particular application. The 3-dB bandwidth specification is helpful, but characteristics such as gain flatness and rolloff characteristic versus frequency must also be determined with the specific capacitive load of the A/D. As a first order analysis, a single pole characteristic can be assumed for the combined effect of the buffer's output impedance and the A/D's input capacitance. If a flat frequency response is required over the signal bandwidth, the attenuation versus frequency is important, as determined from the following equation:

$$A(f) = \frac{1}{\sqrt{[1 + (f/f_3)^2]}}$$

Figure 4-10 demonstrates the gain rolloff which occurs as a function of the 3-dB bandwidth f_3 by converting the equation above to LSBs. Without considering the contribution from bandwidth limitations in the A/D itself, Fig. 4-10 predicts the number of codes which would be lost from the dynamic range of A/Ds with 6, 8, and 10 bits of resolution. The figure dramatically illustrates that to maintain a flat frequency response, buffer amplifiers must have a 3-dB bandwidth, which is at least 10 times greater than the signal bandwidth. At a frequency of only one-half the 3-dB bandwidth of the buffer amplifier, an A/D will have suffered a gain loss of 10% of its codes.

Scaling with VLSI processes can yield the density and switching speed to build flash A/Ds with higher resolution and ultrahigh sampling rates. However, the drive requirements are

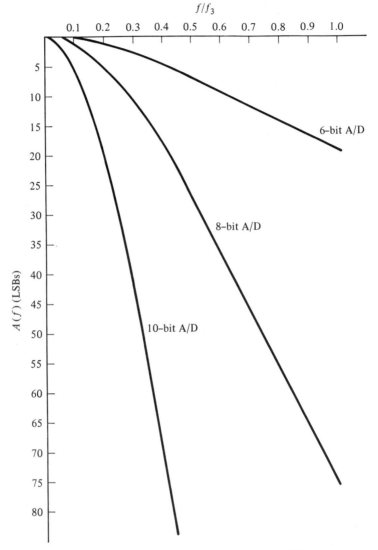

Figure 4-10   Loss in A/D dynamic range as a function of buffer amplifier bandwidth.

not scaled down by the same factor as speed and resolution scale up. The technological advances in the associated linear circuits are more difficult to achieve. T/Hs may become more necessary in order to achieve a comparable gain in usable bandwidth. The AC performance is then controlled by the sampling buffer driving the storage capacitor, with settling response from the driving amplifier becoming the critical parameter for the A/D's performance.

Regardless of the type of driver which is chosen, for signals that approach or in some cases exceed Nyquist frequencies a signal buffer must be able to slew full-scale transitions while settling to within the error tolerance of the A/D. The performance which is achieved will be the result of the combined circuit, which will include all parasitic elements from the buffer output to the A/D input. Such an equivalent circuit is shown in Fig. 4-11. In this RLC circuit, the output resistance of the buffer amplifier drive transistors is in series with stray parasitic inductance from PC board traces, wiring, and internal package pin to IC connections. The output resistance is often specified only at DC, but the actual value will increase at higher signal frequencies.

In T/H applications, the RLC circuit will affect acquisition time and track mode dynamic response. For discrete buffer amplifiers, settling time and stability of high-frequency signals, especially pulses, will be impacted. Because performance in each application depends on a unique combination of parameters, manufacturers will often specify a range for additional series

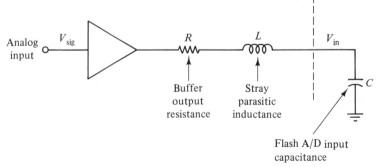

Figure 4-11   Equivalent RLC circuit for input signal path to flash A/D.

resistance to be added in order to "tune" the equivalent circuit for optimum performance. The following analysis will demonstrate how these values are determined. For a voltage step at the buffer input, the response to the A/D input is governed by the following simple equation:

$$V_{sig} = Ri + L\frac{di}{dt} + \frac{1}{C}\int i\, dt$$

Solving for the current using Laplace transform techniques, with initial conditions set to zero, the equation becomes:

$$I(s) = \frac{V_{sig}}{L\,[s^2 + (R/L)s + (1/LC)]}$$

This quadratic equation has the roots

$$b = -\frac{R}{2L} \pm \sqrt{(R/2L)^2 - 1/LC}$$

The boundary condition for stability is for the roots to be real, that is, the square root must be positive: $(R/2L)^2 \geq 1/LC$. From this condition the critical resistance R_{crit} which must be used is defined where

$$R_{crit} = 2 \cdot \sqrt{(L/C)}$$

For the condition of $R < R_{crit}$, the circuit is described as being underdamped, and the input to the A/D will have oscillations which decay exponentially. The step response is optimized when $R = R_{crit}$. For $R > R_{crit}$, the circuit is stable but overdamped. These three cases are illustrated in Fig. 4-12. The upper set of waveforms represents the input signals across a capacitive load which result from a 3-V step, while the lower waveforms illustrate the magnitude of the error versus time, which is $V_{sig} - V_{in}$. Three cases were generated from circuits with $R = R_{crit}/5$, R_{crit}, and $R = 5 \cdot R_{crit}$. For the underdamped case, it can be seen that the peak error is minimized, but there is overshoot and ringing which result in an extended settling time. The peak error is greater in the critically damped case due to a lower slew rate, but the signal settles fastest with a monotonic exponential character-

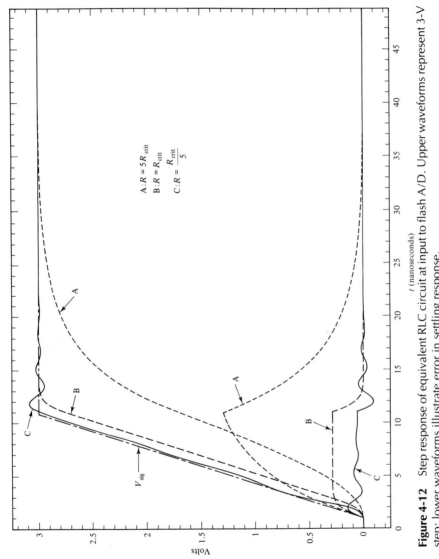

Figure 4-12 Step response of equivalent RLC circuit at input to flash A/D. Upper waveforms represent 3-V step; lower waveforms illustrate error in settling response.

$A: R = 5R_{crit}$
$B: R = R_{crit}$
$C: R = \dfrac{R_{crit}}{5}$

V_{sig}

t (nanoseconds)

Volts

istic. The overdamped case just has longer exponential time constants, which greatly reduce the slew rate.

The real situation of applying the analysis above to the idiosyncrasies of flash A/Ds is more complicated. Since the input capacitance of bipolar devices will vary nonlinearly with the step voltage, the optimum damping resistance will depend on input conditions. Since R_{crit} is inversely proportional to C_{in}, compensation to guarantee stability with the minimum limit of C_{in} would result in overdamping for other portions of the signal range. The overdamping will be most severe at the point where C_{in} is maximum, which will usually be at the limits of the A/D's range. This will compound the problem of loss of dynamic range at high frequencies in bipolar flash A/Ds and limit slewing of pulse inputs.

CMOS A/Ds with switched capacitor inputs present a much different input load, which was described in Chapter 2. The sampling nature of the input can be used to advantage if it is properly understood. The CMOS A/D has two distinct capacitances—one while it is autozeroing and another when it is sampling the input. Full-scale transitions on the input which occur while the sampling switch is off will not be transmitted to the comparators. What this means is that depending on when a transition occurs, pulses can look exactly like DC signals to the CMOS A/D regardless of whether a T/H is used or not. As long as the comparators have been autozeroed, there will be no effects of overload recovery time, slew limiting, or bandwidth limitations.

To optimize the drive circuit to a CMOS flash A/D, the most critical concern is the impulse response of the RLC network when the coupling capacitors of the comparators are switched back to the input signal. The worst-case conditions are for input signals at the extremes of the A/D's range, where all the capacitors must be charged in the same direction in order to acquire the new signal level. If series resistance must be added to optimize this settling time, a smaller R value will be chosen to correspond with the larger input capacitance during sampling. If underdamping occurs, it will primarily affect the signal at the A/D input before it is being sampled, which has no effect on the digitized result. Figure 4-13 illustrates this situation with an example.

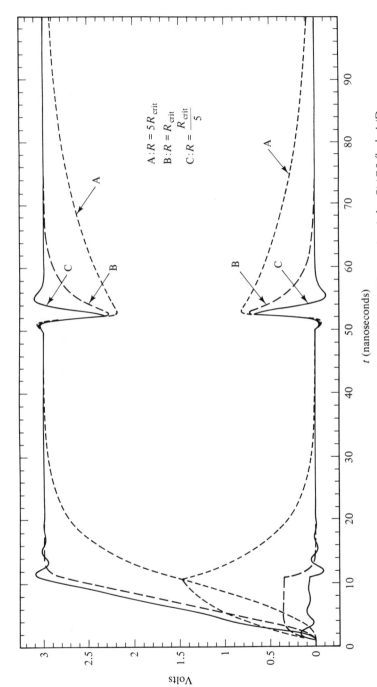

Figure 4-13 Sampling impulse response of equivalent RLC input circuit for CMOS flash A/Ds.

In Fig. 4-13 the same 5 : 1 ratios in R are used as in the previous example, with R_{crit} initially set based on the value of input capacitance during autozeroing. The step from 0 to 3 V is settled in all three cases prior to the sampling impulse (at $t = 50$ nsec.), when the capacitors are switched back to the input. While the circuit with R_{crit} settles fastest after the original slewing, the illustration shows that the lower resistance results in the fastest impulse response during the critical sampling period.

Regardless of the type of A/D, dynamic response will be improved by establishing the smallest possible combination of inductance and capacitance. For this reason, the distance from the buffer to the A/D input should always be minimized. Buffer amplifiers should have low output impedances, which can be compensated by external resistors if necessary. Parasitics can be minimized by using surface mounted ceramic chip capacitors and chip resistors close to the device pins. Sockets should be avoided, but if necessary single pin flush-mounted sockets are recommended. At the expense of higher power dissipation in the buffer, and possibly lower bandwidth and linearity as well, a shunt resistor from the input to ground can benefit both bipolar and CMOS flash A/Ds. Since many high-speed buffers are capable of driving 50-Ω loads, such a low resistance across the A/D's input can improve dynamic performance by reducing the effects of the capacitive load.

Output Demultiplexing

With flash A/Ds now available which are capable of conversion speeds of 300 megasamples per second and beyond, it can be a difficult task for the system designer to deal with the extremely high data rates generated by such devices. Most applications will require that the sampled data be buffered so that it can be processed at the slower clock rate of the signal processing hardware. This can be accomplished by demultiplexing, so that a data rate is achieved which is compatible with available cache memories.

Demultiplexing involves making the data stream wider by splitting it into a number of parallel paths so that the continuous throughput rate is reduced. The data rate from the A/D is reduced by the number of parallel branches N since each path is updated sequentially with every Nth sample. Many of the ultrafast A/Ds implement the simplest form of this scheme by providing an internal 1 : 2 demultiplexer with a dual set of data outputs. Data is pipelined by dividing the sampling clock by two, with the first sample held internally while the second is being processed. Two consecutive samples are then strobed at the output simultaneously. An example of such an architecture is shown in Fig. 4-14.

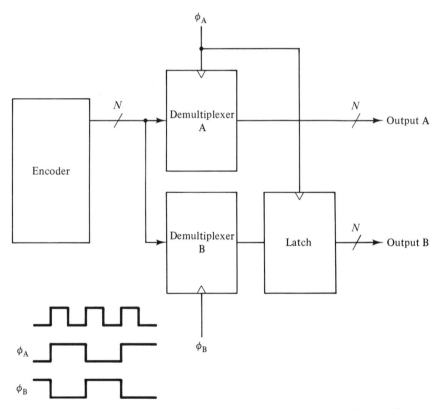

Figure 4-14   Built-in 1 : 2 demultiplexing reduces data rate by a factor of 2 in ultra-fast flash A/D.

The demultiplexing concept can be extended to higher ratios by dividing down the clock further and using more stages of external shift registers to move the data into parallel latches. This technique has been incorporated onto a single 100-MHz ECL chip by Siemens in the SDA 8020. The device, which is referred to as a data acquisition shift register, can also be cascaded for higher demultiplexing ratios.

Complete, Single-Package Flash A/D Solutions

Each of the issues described in this chapter will require the talents of the analog design engineer to develop and test customized solutions for each A/D application. To alleviate some of this effort it may be beneficial in many situations to select a pre-packaged device in which the manufacturer has selected the necessary support components that will optimize performance. Hybrid microcircuit manufacturing techniques have been applied to accomplish this goal by several companies.

Figure 4-15 illustrates the block diagram of one such device, the MN5820 from Micro Networks. Contained within a single 24-pin DIP package is an 8-bit flash A/D and all of the required support functions. A unity gain buffer amplifier for the A/D input signal is left unconnected as an option to the user. An internal bandgap reference circuit and associated amplifiers provide the drive voltages for the resistor ladder. Also included on board is a 75-Ω resistor, which can eliminate the need for external termination of the input signal. The application of power supplies and an external clock is all that remains to digitize high-frequency input signals.

Besides the benefit of reduced board space, reliability also improves with each component attached to a common substrate in a hermetic package. Performance is also enhanced by eliminating the parasitic interconnect between packages. In some hybrid devices the specifications of the individual flash A/D are actually improved by including adjustments for offset, gain, and linearity errors.

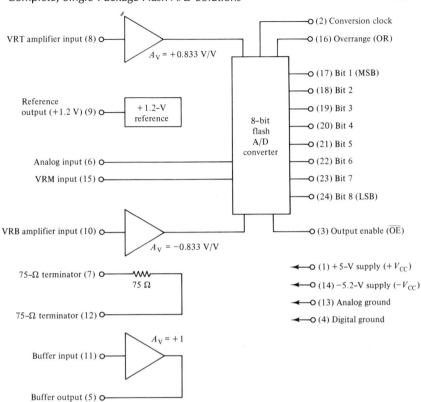

Figure 4-15 The MN5820, a complete, single-package flash A/D, including support circuits. (Reproduced courtesy of Micro Networks, division of Unitrode Corporation, Worcester, MA.)

5 /////////////////

Flash A/D Applications

While flash A/D converters are suitable as is for many traditional low to medium resolution applications, the parallel architectures offer a degree of flexibility which can be exploited in a number of interesting ways. Depending on the priorities of a given design, a flash A/D can be employed to achieve almost any combination of objectives—low power dissipation, extremely high speed, or high resolution. This chapter will describe some of the unique ways in which flash A/Ds can be used.

Burst Mode Sampling

CMOS flash A/Ds that use the autozeroed comparator described in Chapter 2 have several unique application advantages, which are not readily duplicated by devices that use the standard differential comparator. The autozeroing and sampling characteristic of the switched-capacitor circuits gives the user control over the duty cycle and, hence, power dissipation of the CMOS flash A/D. If continuous sampling of the input signal is not required, operation with burst-mode timing contributes further to the low power advantages of CMOS circuits.

As shown in Fig. 5-1, the advantages of burst-mode timing are obtained by returning the A/D's clock to the state which removes the comparators from the autozeroing mode. Static power dissi-

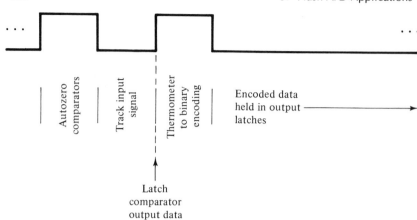

Figure 5-1   Timing diagram for burst-mode sampling in CMOS flash A/Ds.

pation is maximum during autozeroing since this is the only condition where current flows through both the NMOS and PMOS transistors in all of the A/D's comparators. During the input tracking period, power dissipation will vary for each comparator, according to the dynamic conditions of the signal, as the various reference levels are crossed. To minimize power dissipation during idle periods, all of the comparators can be forced to static CMOS logic levels, where no current is drawn from the supplies, by clamping the input signal to ground. This may be necessary for some devices to eliminate drift in the comparators resulting from noise at the input.

Because of the pipelined operation of the A/D, the data at the output will be held during the idle period to the last acquired code until sampling resumes with the next autozeroing clock phase. For a single sample taken with burst-mode timing, the point at which this data is available will depend on the internal architecture of the flash A/D. Some devices will allow the data to propagate directly to the output during the encoding process. Other devices will add another half clock cycle to the pipeline delay, thus stabilizing the data internally before presenting it to the output. When sampling is resumed the first new code output is discarded in order to clear the pipeline.

One-Shot Sampling

Complementary to the burst-mode operation of CMOS flash A/Ds is the one-shot sampling mode. This technique can only be used in devices which lack output latches or allow the output latches to be set to the transparent mode, such as the MN5902 from Micro Networks. In one-shot operation, the A/D is held in the autozeroing mode in preparation for sampling the input. This mode trades the higher continuous power dissipation in the comparators for the advantage of acquiring a digital sample with only one pulse of the clock. Figure 5-2 illustrates the timing used for single-shot sampling with a CMOS flash A/D.

The one-shot mode allows the encoding process to propagate data directly to the output from the comparator latches. When selecting A/Ds with this capability it is vital that the comparator latches have low error rates so that metastable states are not propagated, which may result in sparkle codes at the output. A master/slave latch structure, as shown in Fig. 5-3, reduces the probability of metastable states. By delaying the clock to the

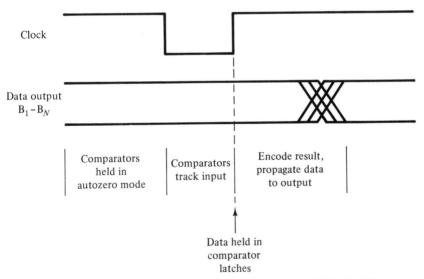

Figure 5-2   Timing diagram for one-shot mode in CMOS flash A/Ds.

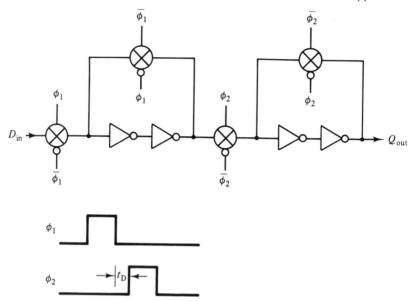

Figure 5-3 CMOS master/slave latch with delayed clocks reduces propagation of metastable states.

slave latch, more time is allowed for the master latch to settle to a stable logic level.

Although the overall throughput time from sample to output is minimized with the one-shot approach, the synchronization advantage provided by the output latches is bypassed. Any internal timing skew between bits in the encoder, which may be especially evident at the MSB transitions, will be propagated to the output. The digital propagation and transition time must be taken into account before further processing is performed.

Stacking Flash A/Ds to Double Resolution

Many flash A/Ds which provide an overflow output signal can be configured in a stacked configuration to double the resolution of the conversion process. An example circuit of this application for a CMOS flash A/D is shown in Fig. 5-4.

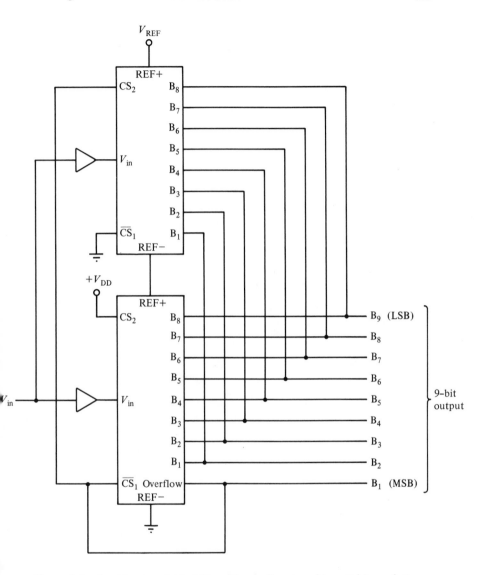

Figure 5-4 Stacking two 8-bit A/Ds with overflow to achieve 9-bit resolution.

By stacking two flash A/Ds, the reference resistor ladders are connected in series. The top of the lower ladder (REF+ lower) is connected to the bottom of the upper ladder (REF− upper). The reference voltage supply, V_{REF} which still must not exceed the specified maximum of a single A/D, is connected from the top of the upper ladder to the bottom of the lower ladder to set the input range of the overall conversion process. The junction between upper and lower A/Ds then becomes the midpoint of the converter.

In a flash A/D only $2^N - 1$ comparators are required to produce the 2^N quantization levels. The all zeros condition, or underflow, is detected when the input signal is less than the first reference voltage. Ideally, this first threshold is set to $\frac{1}{2}$ LSB, which is equivalent to $\frac{1}{2} \cdot V_{REF}/2^N$ or $V_{REF}/2^{N+1}$. The remaining codes 1 through $2^N - 1$ are produced as the input exceeds the individual reference voltages, which increase incrementally until the level of $V_{REF} - 1\frac{1}{2}$ LSBs is exceeded at the last comparator. To produce an overflow signal, an additional comparator is placed 1 LSB above the last comparator at a reference voltage of $V_{REF} - \frac{1}{2}$ LSBs.

In the stacked configuration, the overflow comparator of the lower A/D will produce a logic 1 output when the crossover of the midpoint reference voltage occurs. This will represent the MSB of the stacked converter which properly has a weight of $2^{(N-1)}$. Before this threshold is reached, the data bits are taken from the lower A/D, with the MSB produced by the overflow comparator set to logic 0. For flash A/Ds with the chip select inputs and three-state outputs as shown in Fig. 5-4, the overflow is also used to disable the output from the upper A/D through CS_2 while enabling the lower bits through $\overline{CS}_1$. When the overflow signal switches high, the selection of lower bits switches to the upper A/D.

Since the function of $\overline{CS}_1$ is to only control the data bits and not the overflow output of the lower A/D, the LSBs will properly convert to all zeros at midscale since the upper A/D is still in an underflow condition at this point. Because overflow typically causes the output code from a flash A/D to hold at full scale, designers should be aware that the switching delay on the transition to the upper A/D may momentarily produce the following code transition sequence: $2^{(N-1)}-1 \rightarrow 2^N-1 \rightarrow 2^{(N-1)}$. For example, the midpoint crossover in a 9-bit configuration may produce

$$011111111 \rightarrow 111111111 \rightarrow 100000000$$

The timing delay from the chip select inputs ($\overline{CS}_1$ and CS_2) to the outputs must be added to the normal data propagation delay time from the clock.

Some bipolar flash A/Ds which provide ECL output drivers can also be used in stacked configurations. Unlike CMOS designs, this feature usually requires the manufacturer to produce two versions of a device in order to allow the outputs of the stacked A/Ds to be wire-ORed together. An example of this application using the MN5903 and MN5903A 6-bit flash A/Ds from Micro Networks is shown in Fig. 5-5.

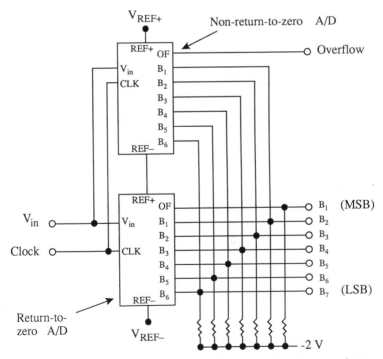

Figure 5-5   Stacking complementary ECL flash A/Ds for higher resolution.

The difference in operation of the two A/Ds illustrated in Fig. 5-5 involves the treatment of the overflow condition. For the lower 6-bit converter in the figure, the sequence of output codes as the input signal approaches overflow is

Over	B_1	B_2	B_3	B_4	B_5	B_6
1	0	0	0	0	0	0
0	1	1	1	1	1	1
0	1	1	1	1	1	0
0	1	1	1	1	0	1

V_{in}

For the upper A/D a more typical counting sequence is employed with a non-return-to-zero format that holds the output bits at full scale:

Over	B_1	B_2	B_3	B_4	B_5	B_6
1	1	1	1	1	1	1
0	1	1	1	1	1	1
0	1	1	1	1	1	0
0	1	1	1	1	0	1

V_{in}

ECL output drivers are typically emitter follower pull-up transistors, which require external pull-down resistors. As in the CMOS example, the overflow signal of the lower flash A/D in Fig. 5-5 acts as the MSB of the composite 7-bit converter. With the six data bits of the two devices connected together, as the input signal exceeds the midscale voltage the lower outputs will hold in the all-zeros condition. This leaves the output drivers of the upper A/D to determine if a logic 1 is generated by turning on a pull-up transistor.

In any stacked configuration, the connection of the two ladders would ideally produce a 1-LSB span for the new midpoint code. However, gain and offset errors in the two converters will combine to determine the voltage span between onset of overflow in the lower A/D and the first code transition in the upper A/D. By corrupting the threshold points, differential linearity errors result at code $2^{(N-1)}$. This error can be reduced by independent adjustment of the voltage to REF− upper and REF+ lower.

In a stacked configuration the user should not expect to achieve exactly twice the resolution of a single flash A/D. Since each individual A/D is operating with half of the total reference voltage span, the size of 1 LSB is reduced by half also. Linearity will inevitably degrade since the noise levels and static errors remain constant. The signal bandwidth and settling response will also be affected if the increase in load capacitance is not taken into account by duplicating the signal path to the two converters.

Piecewise and Nonlinear Transfer Characteristics

Internally, a flash A/D of a given resolution can be decomposed into a stacked architecture as described earlier. For example, an 8-bit A/D is equivalent to four stacked 6-bit A/Ds, each occupying one-fourth of the total dynamic range. In fact, the actual physical layout of the integrated circuit is usually implemented in exactly this way, with the comparators and resistor ladder divided into four equal segments. Because of the physical layout, the intermediate reference voltage taps where the ladder is segmented into halves, quarters, or eighths are usually available externally to the user. In typical applications it is recommended that additional decoupling capacitance be added to these points to improve performance by reducing noise on the reference ladder.

However, access to the intermediate reference taps also allows users to modify the transfer characteristic of the converter. One simple benefit of this feature is the ability to improve upon the inherent integral linearity of the A/D, as discussed in Chapter 4. Small adjustments at the intermediate taps modify the end points of the stacked subranges, nulling out internal errors and bringing the transition voltages of the MSBs back onto the desired straight line through the converter's end points. In other applications the creation of a deliberately nonlinear or piecewise linear transfer characteristic can also yield improvements in performance.

Recall now the calculation of the theoretical limit of an A/D's RMS signal-to-noise ratio, as was derived in Chapter 3:

$$\text{SNR} = 20 \cdot \log \frac{V_{\text{in, RMS}}}{q/\sqrt{12}}$$

The specification as it is usually reported by manufacturers compares the actual result of digitizing a full-scale sine wave to the best-case limit derived from this equation, which is $(6N + 1.76)$ dBs. However, in actual use it is rare that only full-scale signals are being applied to the converter. It is more important in many applications to be able to resolve low-level signals that are imbedded in a much larger total dynamic range. In such cases the intermediate reference taps can be used to implement a piecewise linear or companding (compression and expansion) transfer characteristic to improve SNR over important segments of the A/D's range.

In Fig. 5-6 the solid straight line demonstrates the limit on SNR which can be achieved with a linear transfer characteristic versus the peak value of a sine-wave input signal. By considering the flash A/D segments as stacked A/Ds, it should be clear that 6 dBs (or 1 bit) of SNR is lost for signals which are digitized entirely below the midscale of the A/D. If signals fall entirely in the first quarter of the A/D's range, 12 dBs (or 2 bits) are lost from the theoretical full-scale limit. These results can be derived from the equation above by reducing the signal amplitude while the quantization noise, or LSB size, remains constant. The key then is to reduce the quanta size q to achieve more resolution in the range of interest.

The dashed line in Fig. 5-6 demonstrates the result of *reducing* the midscale reference voltage of a flash A/D so that it is set to $\frac{1}{4}$ full scale rather than $\frac{1}{2}$. This will double the number of quantization steps that are applied to signals within the first quarter of the A/D's range, potentially increasing SNR for such signals by as much as 6 dBs. High-level signals are compressed into fewer codes, while low-level signals are expanded into a wider range. The modified quantization noise characteristic is illustrated in Fig. 5-7.

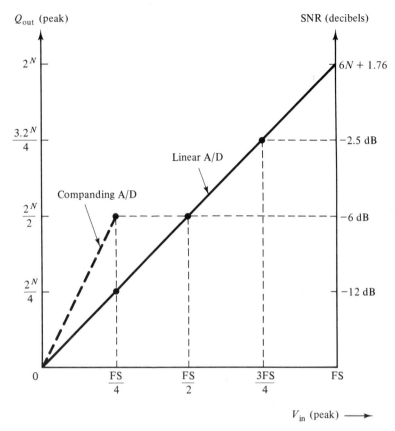

Figure 5-6 Reduction of midscale voltage improves resolution of low-level signals.

In the example of the piecewise linear approach represented in Fig. 5-7, half of the A/D's codes are 50% shorter than in the ideal straight-line converter, and half of the codes are 50% longer. A similar case was analyzed in Chapter 3 to demonstrate the effects of differential nonlinearity on SNR. Recall from that analysis that, when compared to an ideal uniform code width, the short codes were shown to produce 50% less noise than nominal while the long codes produced 50% more noise. The difference in this application is that the probabilistic weighting which is used to

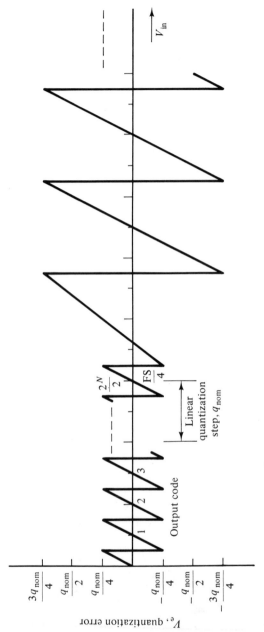

Figure 5-7 Quantization noise in piecewise linear A/D.

determine overall RMS noise is different, that is, the application has determined that the "short" codes will occur more frequently than in the case of the straight-line converter. As an example, if the signals being digitized have a distribution such that 50% of the samples are in the first quarter of the A/D's range and 50% are in the last three quarters,

$$v_{e,RMS} = \frac{q_{nom}}{\sqrt{12}} \cdot \sqrt{.5 \cdot (.5)^2 + .5 \cdot (1.5)^2}$$

$$= 1.12 \cdot \frac{q_{nom}}{\sqrt{12}}$$

The 12% increase in the RMS noise in this example implies that the average SNR from digitizing signals which follow the expected distribution decreases by approximately 1 dB compared to a linear approach; but for the critical low-level signals SNR increases by up to 6 dB. However, a further price which is paid for this increase in low-level performance is an equivalent loss of SNR for high-level signals, that is, those entirely within the upper range of the A/D. These signals are digitized completely with the longer codes, so that the lower resolution results in a 6-dB loss of SNR compared to a linear A/D.

With the companding approach, each segment of the flash A/D's range which is accessible to the user can be customized to shape the noise characteristic of the conversion process. The same caveat used for the stacked architecture also applies here: in attempting to increase resolution static linearity errors will limit the actual gain in resolution which can be realized.

Oversampling for Increased Effective Resolution

The analyses of quantization noise effects that have been described up to this point have been based on the effective RMS voltage of the characteristic sawtooth error signal in the time domain. A uniform quantization step which is equivalent to the sawtooth amplitude is assumed for the ideal A/D. This results in a uniform probability distribution function for RMS noise versus input voltage. Similar assumptions are also used for the idealized model in the frequency domain, resulting in a constant power spectrum versus input frequency.

The RMS power in the noise signal will be constant regardless of the sampling rate and Nyquist limits. This fact can be exploited in oversampling A/D converters. A brief review of some of the background concepts will aid in the explanation.

In the time domain, the energy of a signal is

$$E = \int_{-\infty}^{\infty} [x(t)]^2 \, dt$$

This energy could also be expressed in the frequency domain, according to Parseval's theorem, where $|X(j\omega)|^2$ is referred to as the energy spectrum:

$$E = \frac{1}{2\pi} \int_{-\infty}^{\infty} |X(j\omega)|^2 \, d\omega$$

In a similar fashion, the expressions above can be used to relate the average power of a signal to its power spectral density $S_X(\omega)$:

$$\overline{x^2} = \lim_{T \to \infty} \frac{1}{2T} \int_{-\infty}^{\infty} [x(t)]^2 \, dt = \frac{1}{2\pi} \int_{-\infty}^{\infty} S_x(\omega) \, d\omega$$

The analysis of A/D signal-to-noise ratio in Chapter 3 showed that the average power of the quantization noise is $q^2/12$. Due to the limits of sampling theory, in the frequency domain the spectrum is constrained to the frequencies between $\pm F_s/2$. This characteristic is represented in Figure 5-8. By definition, in an oversampling A/D the bandwidth of the input signal is con-

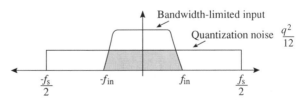

Figure 5-8 Oversampling reduces the quantization noise power contained within the input signal bandwidth.

strained to be much less than the Nyquist limit. In such cases the quantization noise spectrum will extend beyond the spectrum of the input signal. The ramification of this is that the total quantization noise power in the bandwidth of interest is smaller; in fact, it will be inversely proportional to the oversampling ratio:

$$\text{Oversampling ratio} = \frac{f_{\text{NYQUIST}}}{f_{\text{in}}} = \frac{f_s}{2 \cdot f_{\text{in}}}$$

In digital signal processing applications, a digital low-pass filter at the A/D output could be used to effectively remove the quantization noise which is contained in the unwanted part of the Nyquist bandwidth. This filter will only pass the input signal bandwidth with its proportionately reduced noise power, resulting in an increased SNR:

$$\text{SNR} = 10 \log \left[\frac{(2^N q / 2\sqrt{2})^2}{q^2/12} \cdot \frac{f_s}{2 \cdot f_{\text{in}}} \right]$$

$$\text{SNR} = 6.02N + 1.76 + 10 \log \left[\frac{f_s}{2 \cdot f_{\text{in}}} \right]$$

The equation above shows that for each doubling of the sampling rate relative to the input bandwidth, the theoretical limit on SNR can be increased by 3 dBs, an increase in effective resolution of $\frac{1}{2}$ bit. This technique is the key to delta-sigma oversampling A/Ds, which are presently available with up to 20 bits of effective resolution when processing audio band signals.

Reference Multiplication for Mixing and Gain Control

Although designers in most applications strive to maintain stable DC reference levels to an A/D, the floating terminals of the resistor ladder in a flash A/D create the potential to dynamically adjust the range of the converter. This feature can be exploited in several unique ways.

The allowable range of adjustment for reference voltages will depend on the manufacturer's specifications, but CMOS devices usually have more flexibility than their bipolar counterparts. This difference is due to the DC coupling of the signal path and limited common-mode range for the comparators in the bipolar A/D. The AC coupling of the autozeroed CMOS comparator provides more latitude in setting the reference levels.

To better understand the effect of dynamically adjusted reference levels, the equation below will represent an ideal linear transfer function of an A/D converter (offset of the first code transition is ignored for purposes of this discussion):

$$Q_{out} = INT \left[\frac{V_{in}(t)}{V_{REF}(t)} \cdot 2^N \right], \quad 0 \leq V_{in} < V_{REF}$$

In most discussions an integer value for the output code is implied as shown above, effectively placing the "decimal" point (really a binary point in this case) to the right of the LSB. However, the output code Q_{out} actually represents the value in binary code of the fraction formed by normalizing the input signal by the full-scale reference voltage, with the result truncated by the N-bit resolution of the A/D. For example, the binary output code 10110_2 actually represents an input signal of $0.10110_2 \cdot V_{REF}$. This interpretation of the A/D function leads to a new application, which is to implement an N-bit divider for two analog signals with the function of

$$Q_{out} = V_{in} \div V_{REF}, \text{ for } V_{REF}(t) > V_{in}(t) > REF-$$

From the ideal linear transfer function, the gain of the converter, which is the slope of the straight line through the A/D's end points, is found to be

$$Gain = \frac{Q_{out}}{V_{in}} = \frac{2^N}{V_{REF}}$$

If the reference voltage is adjusted dynamically, the A/D's transfer function becomes

$$Q_{out}(t) = INT [G(t) \cdot V_{in}(t)]$$

The implementation of AGC (automatic gain control) functions, which are sometimes performed on the analog input preceding the A/D, can be made an integral part of the conversion process by continuous adjustment of V_{REF}. Application of an AC signal to the flash A/D reference may also be thought of as amplitude modulation of the signal. However, in such cases the reciprocal dependence on the V_{REF} level must be kept in mind. It may not be immediately apparent that linear modulation of the A/D's reference will produce a nonlinear modulation effect in digitizing the analog input. This nonlinear gain characteristic is illustrated in Fig. 5-9, where the reference is varied symmetrically by ±50%.

Bandwidth of the reference signal path must be considered in any application where an AC signal will be applied. Of course, all high-frequency decoupling capacitors which are normally used on the reference ladder must be removed. The analysis is somewhat different for the two types of flash A/D comparators. The same parameters which limit the input signal bandwidth to the differential comparator will have an even greater effect on AC signals applied to the resistor divider reference circuit.

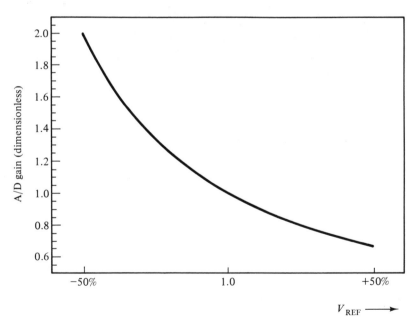

Figure 5-9 Nonlinear reciprocal relationship of flash A/D gain versus V_{REF}.

In a bipolar flash converter the combination of series ladder resistance together with the input capacitance from the emitter-follower buffer stages forms a distributed RC filter. Since the ladder resistance is typically on the order of 100 Ω, a substantially lower bandwidth can be expected from the reference signal path than from the input signal path. It is also important to remember that the variation of junction capacitance versus junction bias along the ladder is nonlinear, which will cause distortion of any AC signals present on the resistor ladder.

In CMOS flash A/Ds the distributed RC filtering still limits reference bandwidth, but the input capacitors have little or no voltage coefficient to cause nonlinearity. However, a different effect takes place due to the condition of the comparators during the autozero cycle. In the differential pair circuit the reference is continuously compared to the input signal during the sampling interval. Autozeroing architectures utilize sequential sampling of the reference and input. AC signals on the reference will be sampled and held on one cycle, setting the threshold for comparison to the input on the next cycle. Reference adjustments cannot be made simultaneous to the sampling of the signal input.

During the application of autozeroing feedback, the bias potential of the CMOS comparator may be altered from its nominal switching threshold if the amplitude of the AC voltage imposed on its input becomes large. The signal path through the input coupling capacitor creates a high-pass filter, which can cause the performance of the comparators to vary with the frequency and sampling point of the reference waveform. This effect will set the upper limit on the reference bandwidth in CMOS flash A/Ds. Equivalent circuits for the reference signal paths in bipolar and CMOS A/Ds are compared in Fig. 5-10.

Ping-Pong Configuration to Double Conversion Rate

By connecting two flash A/Ds so that their analog inputs and digital outputs are in parallel, the effective sampling rate may be doubled without actually increasing the clock rate to the A/Ds. This configuration is often referred to as a "ping-pong" architec-

REF+ 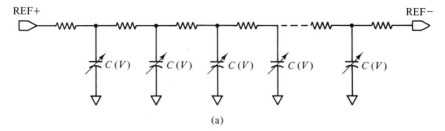 REF−

(a)

REF+ 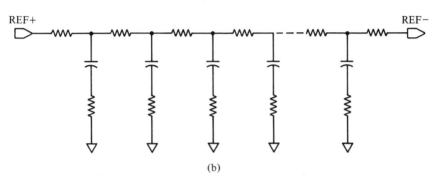 REF−

(b)

Figure 5-10 Equivalent circuit for AC signals applied to reference ladder in flash A/Ds: (a) Bipolar A/D, (b) CMOS A/D.

ture since it is based on the two A/Ds sampling 180° out of phase with each other, resulting in a digital output which "bounces" between A/Ds. A block diagram of a ping-pong A/D is shown in Fig. 5-11.

The circuit of Fig. 5-11 relies on a flash A/D with dual opposite polarity output controls. The sampling clocks are shown as simple inverses of each other; but in critical applications tighter control may be necessary to achieve more accurate 180° phasing and to minimize jitter between the two sampling clocks. A timing diagram for the ping-pong architecture is shown in Fig. 5-12. By directly connecting the output controls to the clock, a multiplexing action occurs on the digital data bus. Each A/D's output will be valid on the bus for one-half of the clock period. The exact window when each byte is valid will include the delay from CS to the output.

An alternative architecture for A/Ds which lack the inverse output controls is to use an external 2 : 1 multiplexer to select the

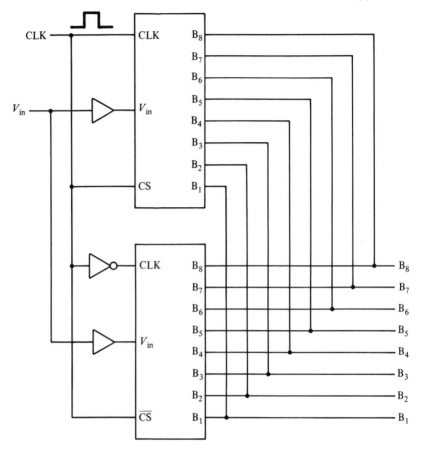

Figure 5-11   Ping-pong architecture to double speed of A/D conversion.

alternating bytes of data. An advantage to such an approach is that the select line clock can be optimized for the proper timing within each valid data period. This will resolve timing problems which could occur in transferring the data for further processing at the 2× rate.

In doubling the sample rate this way, designers must be aware that the ping-pong technique will combine the error which results from the inevitable mismatch of the two A/Ds. Even for a case where the converters both hold to ±½-LSB linearity, the code transition points for the two A/Ds will not necessarily be coincident, as demonstrated in Fig. 5-13a. When the alternate digital outputs are combined in a series of samples at a rate of $2 \cdot F_s$, oscillations between consecutive codes and a loss of precision at the code edges will result as shown in Fig. 5-13b. This undesirable effect will worsen for A/Ds with poorer linearity and matching, resulting in alternate code steps of greater than 1 LSB in some cases.

By doubling the data sample rate, the ping-pong A/D also doubles the effective Nyquist bandwidth of the combined conversion process. However, this increase in bandwidth may not be fully realized, since the bandwidth limitations of the individual A/Ds ultimately limit performance. Since the total input capacitance doubles also, it is best to keep the input signal paths separate to minimize degradation in the buffer amplifier drive capability and bandwidth.

Extensions of the ping-pong architecture to more than two flash A/Ds can also be done to create converters with extremely high effective sampling rates. This technique, sometimes referred to as *interleaving*, is used in applications such as digital sampling oscilloscopes (DSOs). By counting down from a higher fre-

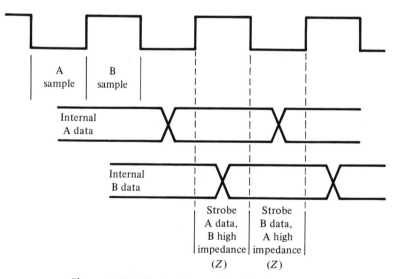

Figure 5-12   Timing diagram for ping-pong A/D.

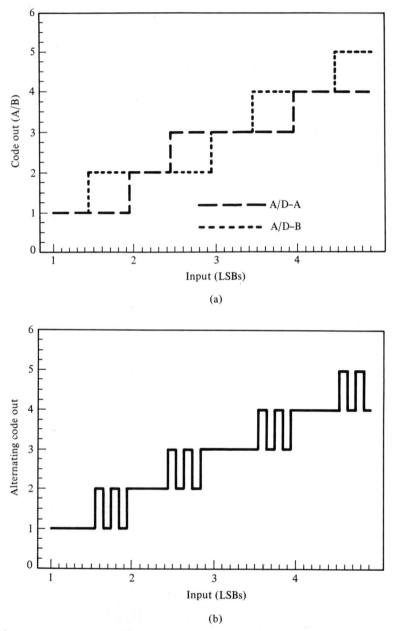

Figure 5-13   (a) Mismatch in quantization steps for ping-pong A/D. (b) Loss of precision in code transitions for ping-pong A/D.

quency clock, or using precision delay lines, the A/Ds can be strobed sequentially at small time intervals. The digital outputs can be combined into a sequential data record by using a fast FIFO (first-in, first-out) memory to store the results.

Subranging A/Ds

A basic description of the subranging or two-pass architecture was covered in Chapter 1. This technique will now be reexamined in more detail by including the error sources which must be considered when constructing such a converter. With an understanding of the limitations involved in extending flash A/Ds to such high resolution applications, the digital compensation techniques which are used to overcome them will become clear.

In actual practice two passes through an N-bit flash A/D can not produce a converter with $2N$ bits of resolution. Quantization of a flash A/D's error through subranging would effectively divide each step of the first-pass (or coarse) conversion into 2^N steps in the second-pass (or fine) conversion. To double the number of bits of resolution in two passes, so that the results can be simply appended together, requires a linearity specification of $\pm \frac{1}{2} \cdot \frac{1}{2^N}$ LSBs (for example, $\pm \frac{1}{128}$ LSB from a 6-bit A/D). This unrealistic requirement would be the maximum allowable error that would still permit the two 6-bit conversions to be combined for a 12-bit output with no missing codes (i.e., DNL $= \pm$ 1 LSB).

As a more realistic example, consider the error which results when using a 7-bit flash A/D to build a 12-bit subranging A/D. If the flash A/D is specified for $\pm \frac{1}{2}$ LSBs of differential and integral linearity, it could have step sizes between codes of $\frac{1}{2}$ to $1\frac{1}{2}$ coarse LSBs, which would then represent the quantization error in the first-pass conversion. To achieve 12 bits of resolution requires accuracy to $\frac{1}{32}$ of an LSB at the 7-bit level ($2^{-12}/2^{-7} = \frac{1}{32}$). If the quantization error in the first pass is $1\frac{1}{2}$ LSBs, this is equivalent to 48 LSBs at the 12-bit level. This error is expanded and quantized in the second-pass conversion, which will require digital correction schemes to combine the two-pass results into a 12-bit output.

It is important to note that in order to properly calculate the first-pass quantization error the linearity specification of the first-pass A/D must be based on a true end-point measurement. Best-fit or code-centered measurements, as discussed in Chapter 3, will not give a true indication of step size in the first-pass conversion.

To produce the signal for the second-pass conversion, the coarse bits are processed by a high-resolution D/A. The function of the D/A is to map the end points of the subrange segments onto the straight line which maintains the desired end-point linearity of the overall A/D. This effectively creates an analog measurement of the coarse N-bit first-pass code transition voltages with an accuracy of $2N$ bits. Any error in these measurements passes directly to the second-pass conversion. This point is demonstrated in Fig. 5-14. Although the D/A only needs the same number of bits as the coarse A/D, it must possess linearity and accuracy which will be consistent with the desired two-pass performance.

In the residue amplifier the end points of the subranges determined by the D/A are subtracted from the input signal, creating a new analog signal which represents a measurement of the first-pass quantization error with $2N$ bits of accuracy. The error amplifier which is used for this measurement must possess linearity that is better than $2N$ bits over the entire input range. This process is the critical foundation of the subranging architecture. Regardless of the magnitude of the first-pass error, it can, theoretically at least, be quantized to whatever resolution is desired by a second-pass A/D conversion. However, if the D/A and error amplifier are nonlinear, the measurements themselves will have linearity error, which will directly add to the result and ultimately limit the subranging A/D's linearity.

In the process of creating the first-pass residue signal, the cumulative first-pass quantization error (S/H + A/D + D/A) must be amplified to fit the range of the second-pass converter. In the simplified architecture with ideal flash A/Ds, two passes through the same converter produce uniform coarse steps of 1 LSB, which are equivalent to exactly $2^{N/2}$ bits in the total conversion. By setting the gain of the error amplifier to $2^{N/2}$, the

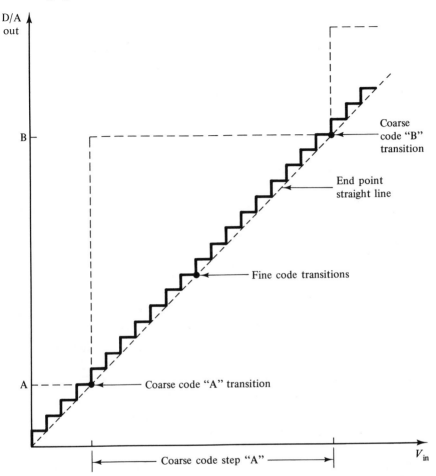

Figure 5-14 Mapping of coarse quantization steps onto straight line with fine resolution.

residue would be expanded back to the full range of the flash A/D. The fine bits N_f are then simply concatenated with the coarse bits N_c to produce the N-bit result.

If different A/Ds are used in the first and second passes, the ideal coarse steps would be $2^{(N-N_c)}$ or 2^{N_f} bits. The first-pass error is then multiplied by 2^{N_f} for expansion to the full range of the second-pass A/D. Returning to the earlier example, using a

7-bit A/D in the first pass would require a 5-bit second-pass conversion, and hence the residue amplifier gain would need to be 2^5 or 32. However, if the 7-bit A/D has linearity of $\pm\frac{1}{2}$ LSBs, the first-pass quantization error of a long code, which is normalized back to full scale by the residue amplifier, would produce an error signal ε_1 with an amplitude of

$$\varepsilon_1 = 2^5 \cdot \frac{1.5 \cdot 2^{N-N_c}}{2^N} = 2^5 \cdot \frac{1.5 \cdot 2^N \cdot 2^{-5}}{2^N} = 1.5 \text{ FS}$$

Obviously, this would cause the second-pass A/D to overflow. A solution might appear to be to cut the residue amplifier gain by a factor of two, which would keep all foreseen first-pass errors within the full-scale range of the second-pass converter. However, this would cause an overall loss of resolution since ideal coarse steps (i.e., where quantization error is 1 LSB) would be compressed to produce a subrange which uses only half of the second-pass A/D and thus produces only 4 bits of information. From a perfect first-pass A/D, the fifth bit (MSB) of the second pass would always be a logic 0, causing step discontinuities of 16 counts (2^4) in the 12-bit transfer characteristic at the subrange boundaries.

It is apparent that reducing the residue amplifier gain must be accompanied by an "extra" bit of resolution that must be added to the second-pass converter, that is, a 7-bit first-pass A/D with a 6-bit second-pass A/D so that 1 bit can be allocated to the overflow condition. Then there will always be 5 bits of quantization available to be combined with the first-pass result.

With positive DNL in the first pass, this process of creating an extra overflow bit leads to a simple digital corrector scheme. Such a circuit is nothing but an adder that adjusts the first-pass MSBs based on the amount of error measured in the second pass:

First pass
MSBs $\rightarrow$ $D_{11}D_{12}D_{13}D_{14}D_{15}D_{16}D_{17}0$ 0   0   0   0
Second pass
LSBs $\rightarrow$ $+0$ 0   0   0   0   0 $D_{21}D_{22}D_{23}D_{24}D_{25}D_{26}$
Corrected
output $\rightarrow$ B_1 B_2 B_3 B_4 B_5 B_6 B_7 B_8 B_9 B_{10} B_{11} B_{12}

When the length of a first-pass code generates greater than 1 coarse LSB of quantization error, the addition of the MSB from the second pass will correct the output code. In effect, the extra bit tells us that the first-pass residue actually crossed into the next subrange on the transfer function, so the first-pass code must be increased by 1 bit.

In actuality, as earlier examples have shown, the first-pass quantization error can be positive or negative. Since most flash A/Ds are unipolar, the case of short first-pass codes must be handled in a slightly different fashion. If the output of the residual amplifier is offset by a fixed amount the error signal can be forced to be unipolar also, preventing the second-pass A/D from under-flowing. To compensate, the digital correction logic can adjust the second-pass code by subtracting the digital equivalent of the residual offset from the final result.

It should be apparent from the discussion above that the design of a high-resolution subranging A/D is complex and may require several custom circuits to fully implement. Fortunately for potential users there are many single-package solutions which are available from hybrid data converter manufacturers. By choosing the optimum components for each function block, developing custom ICs where necessary, and laser trimming for accuracy, these devices can provide a ready-made solution for many high-speed applications. An example of such a component is the MN6249 from Micro Networks, illustrated in the block diagram of Fig. 5-15. This device includes the required T/H circuit along with the core flash A/D, high-resolution D/A, and a custom error correcting logic chip to provide a complete 12-bit, 2-MHz sampling A/D.

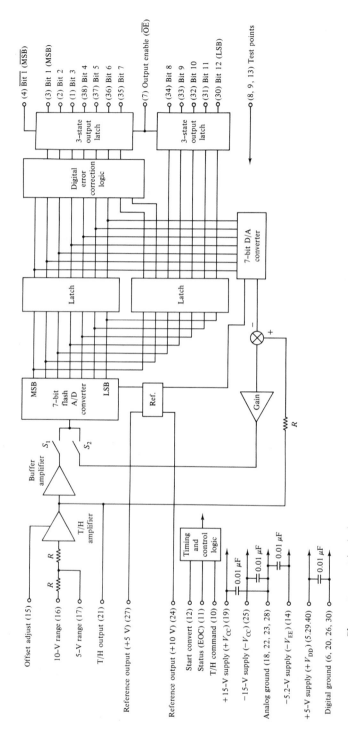

Figure 5-15 Block diagram of MN6249 from Micro Networks, a complete 12-bit, 2-MHz sampling A/D. (Reproduced courtesy of Micro Networks division of Unitrode Corporation, Worcester, MA.)

6 ///////////////////

Test Methods for High-Speed A/D Converters

After a flash A/D has been selected based on a thorough examination of specifications and the support circuitry has been designed and built, the next step is to perform tests to verify that the performance goals have been achieved. This step can be the most difficult of all, and there are many approaches to take, depending on the particular circumstances. Options range from traditional analog bench instruments to PC-based systems and dedicated mixed-signal testers. This chapter will review the types of tests which should be made to verify flash A/Ds and describe reasonable testers which can be assembled in a typical engineering lab. For high-volume applications, turnkey solutions are available from major ATE vendors but at a cost in the range of $1 million.

Static Linearity Tests

Analog Difference Signal Method

The discussion of A/D specifications in Chapter 3 described how differential and integral linearity can be defined on the basis of the sawtooth quantization error plot. If the output of an A/D converter is used to drive the input of a higher accuracy D/A, direct comparison to the original analog input signal can be performed to generate this waveform.

Figure 6-1 shows the block diagram of a test setup to generate a quantization error plot. A triangle waveform which spans the range of the A/D is used as the input signal to the A/D. The gain of the reconversion D/A is adjusted so that its full-scale voltage matches the input range. Subtraction of the quantized output from the analog input can be performed by an op-amp circuit or by applying the two signals to a difference amp plug-in for the vertical drive of an oscilloscope, such as the Tektronix 7A13. The input signal can be used as the horizontal drive to the oscilloscope, which will create the full range error waveform as shown in Fig. 6-1.

To measure differential linearity, a delayed sweep and time-base multiplier of the oscilloscope can be used to scan individual segments of the sawtooth error waveform. By initially calibrating the x-axis to the ideal width of 1 LSB, the variation in width of the sawtooth segments can then be measured with the oscilloscope graticule to find the limits of differential linearity error. For integral linearity the entire series of steps between the end points can be displayed. For an ideal A/D, all of the sawtooth peaks would line up on the same vertical grid point. The extent of the bowing between peaks can be compared to the nominal step height to measure integral linearity. The error signal will go off scale at the overflow and underflow points.

For high-speed A/Ds the difference signal method is useful mostly as a qualitative screening test, since a poor device will reveal weaknesses even under low-frequency conditions. The difficulty of obtaining a high-speed D/A with 2 to 4 bits greater resolution and accuracy than the A/D limits the amount of dynamic testing that can be done with this method. A slower D/A can be used if the data from the A/D is "decimated." This process requires an external latch, which is strobed at a lower rate by a divided down clock, so that one digitized sample is sent to the D/A for some multiple number of samples taken by the A/D.

It can also be very tedious to scan all the codes of the A/D using the difference signal method. Settling time performance from the D/A will affect the peaks of the error waveform, especially at the MSB transitions of the D/A. The input signal should

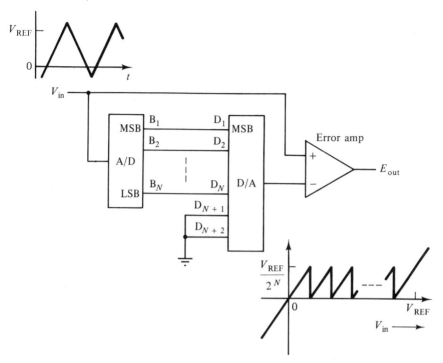

Figure 6-1   Static linearity test with analog difference signal test method.

also have a linearity at least as good as the D/A, so that its contribution to the error waveform shape is not significant.

Crossplots

The crossplot method of measuring static linearity has been widely used for evaluating successive approximation A/D converters. Only common laboratory equipment is required, such as an oscilloscope, pulse generator, and triangle waveform generator. Since measurements depend on a visual examination of oscilloscope traces, this method is suitable for quick evaluations where a high degree of precision is not required.

In the crossplot test a simple D/A conversion of some of the LSBs from the A/D under test (typically, the last 2 bits) is used as the vertical input to the oscilloscope. An input signal to the A/D is derived by summing an accurate DC level from a high-precision

D/A with a triangle waveform from a function generator. This creates a signal which will span several codes of the A/D. The AC waveform is also used to drive the horizontal input to the oscilloscope. The block diagram in Fig. 6-2 describes the test setup.

With a linear signal at the input the A/D should count sequentially through a series of codes, with each output code being mapped to exactly equal voltage increments of the input waveform. By decoding the two LSBs, a staircase waveform with four distinct voltage levels should result. Differential linearity errors will be manifested by unequal step widths in the staircase. Figure 6-3 demonstrates the analog decoding of the LSB counting sequence.

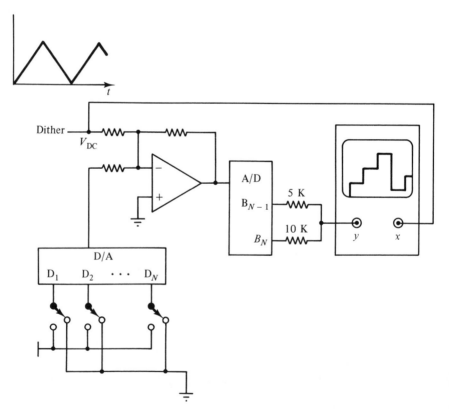

Figure 6-2   Static linearity test with crossplot test method.

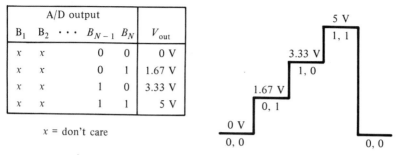

A/D output				V_{out}
B_1	B_2	$\cdots$ B_{N-1}	B_N	
x	x	0	0	0 V
x	x	0	1	1.67 V
x	x	1	0	3.33 V
x	x	1	1	5 V

x = don't care

Figure 6-3   Analog decode of LSBs for crossplot test.

The crossplot test achieved popularity in early A/D testing because of its simplicity and speed, which is based on an assumption that all codes do not have to be checked. In successive approximation A/Ds the internal D/A creates 2^N voltages (or currents) by adding N binary weighted voltage or current sources. If perfect superposition of these sources can be assumed, only the relative weights of the N bits, or the so-called "major carries," need to be measured. This assumption does not hold for flash A/Ds, where the 2^N reference levels are created by 2^N distinct segments of a resistor ladder. The method can still be used, but all of the codes must be checked to guarantee linearity. This makes the crossplot very tedious for flash A/Ds.

Servo-Loop Code Transition Measurement

An analog measurement method which can provide a high degree of accuracy, and also lends itself to being automated, is the servo-loop system shown in Fig. 6-4. Many ATE systems use this technique to measure A/D linearity, but PC-based controllers perform just as well and provide a much more economical alternative.

The loop begins with an operational amplifier which is configured as an integrator. Transistor switches at the integrator input are used to select between positive and negative DC voltage sources. With a constant DC voltage at the integrator input a linear ramp will be generated which is sampled by the A/D converter. A digital comparator has one input supplied externally

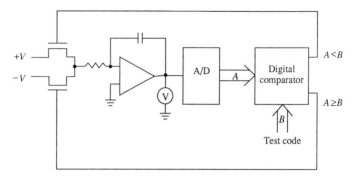

Figure 6-4 Integrating servo-loop tester.

which is used to specify the code transition point to be measured. The second input to the comparator logic comes from the A/D output code. If the A/D output is less than the selected code, the A < B signal will be true, which allows a monotonically increasing ramp to be produced by the integrator. Conversely, when the A/D exceeds the selected code, a monotonically decreasing ramp is produced by the A ≥ B signal.

The test begins by finding the end points of the A/D transfer characteristic. By measuring the first and last transition voltages, the nominal, or average, size of an individual code can be determined. This value is used for the linearity measurements. With code 1 loaded in the comparator, the ramp slews monotonically until it is approximately equal to the A/D's offset voltage. If the ramp approaches the code 1 transition voltage from below, the A < B signal becomes false and A ≥ B becomes true, causing the ramp to reverse direction. Depending on the integration time constant and loop delays, a sequence of samples eventually produces code 0, again causing the ramp signal to increase. The feedback of the servo loop causes this oscillatory behavior to continue, producing a triangle waveform which is centered about the DC level of the code transition. The voltmeter at the integrator's output provides an accurate measurement of the transition voltage. When the loop time constant is properly adjusted, the amplitude of the triangle waveform should ideally be a fraction of an LSB.

A similar procedure is used to measure the full-scale transition point. After the two end points of the A/D's transfer characteristic are found, the nominal LSB is calculated from the following equation:

$$LSB_{nom} = \frac{V_{FS} - V_1}{(2^N - 1) - 1} = \frac{V_{FS} - V_1}{2^N - 2}$$

It should be observed that the inverse of the nominal LSB is equivalent to the slope of the true end-point straight line defined by the A/D's transfer characteristic. A plot of the code output (y-axis) versus V_{in} (x-axis) should exhibit a change by exactly one count for each increment of the input voltage equal to one nominal LSB. The A/D gain $dCODE_{out}/dV_{in}$ is determined by this slope. This information can be used to determine gain error by comparison to the ideal transfer characteristic:

$$GAIN_{ideal} = \frac{1}{LSB_{ideal}} = \frac{2^N}{V_{REF+} - V_{REF-}}$$

The complete servo-loop test proceeds by recording the transition voltages of each code in a high-speed A/D. Averaging may be employed to increase the accuracy of the measurements and reduce noise effects. From the recorded transition voltages, DNL and INL are calculated as shown below:

$$DNL_i = \frac{V_{i+1} - V_i}{LSB_{nom}} - 1 \text{ (LSBs)}$$

$$INL_i = \frac{V_i}{i \cdot LSB_{nom} + V_1} - 1 \text{ (LSBs)}$$

Problems can arise in the servo-loop test when an A/D exhibits hysteresis-like behavior. An A/D's comparators may not slew or settle symmetrically with small overdrive voltages, causing a response which depends on the direction of approach to the code edge. Digital noise feedback and internal D/A settling near switching points can also be code and direction dependent, which can cause a shift in the measured transition voltage. In extreme cases it may be found that most or all of the span of an individual A/D quantization step exhibits noise resulting in alternating transitions between consecutive codes, making an exact determination of the transition voltage impossible. These errors are essentially a measure of precision, stability, and added noise, which, as was mentioned briefly in Chapter 3, are never specified by manufacturers. Factors such as these may cause the dither signal to oscillate nonuniformly with an amplitude greater than an LSB, making an accurate measurement of linearity difficult to obtain with the servo-loop approach. Figure 6-5 compares the dithering triangular error waveform from an ideal A/D to a more realistic signal that is affected by noise in the conversion process. Digital techniques can be used to overcome these problems by analyzing the probability of each code over the range of noisy transitions.

To minimize the propagation of digital noise in the test fixture, with some A/Ds it may be beneficial to buffer the input to the digital comparator with a separate latch. With proper timing this may allow the A/D output drivers either to be held in the high-impedance state during the sampling clock phase or to suppress transients which may result in feedback to the input signal sources.

In automated applications it may be beneficial to provide a mechanism for decreasing the integration time constant in order to reduce the wait time for full-scale slewing of the input signal. A high-resolution D/A can also be used to control the input voltage to the A/D, either in combination with the integrator as in Fig. 6-4 or alone as in Fig. 6-6. Code transitions can be determined di-

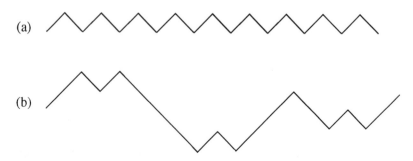

Figure 6-5 Dithering integrator output: (a) Ideal noiseless code transition; (b) A/D code with noise and hysteresis.

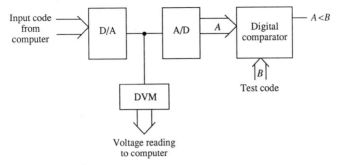

Figure 6-6 D/A-based sevo-loop test.

rectly in software by acquiring the output byte from the A/D or by monitoring the output of the digital comparator. The D/A-based approach allows the input voltage to slew the full range of the A/D in much less time than the integrator ramp. By providing full digital control, the tester hardware can be made universal to any A/D, eliminating adjustments in RC time constants. With a coarse and fine D/A, as in Fig. 6-7, resolution of the input signal can be obtained that is higher than that of the individual D/As. This approach may be necessary for higher resolution A/Ds, but integral linearity of the input ramp is ultimately limited by the D/A's performance. Integral linearity of the source should be 3 to 4 bits better than the A/D that is being measured.

Digital Analysis of Static Linearity

An efficient method of measuring the static linearity of a high-speed A/D is to use a computer to measure error directly from the digital data output of the converter. A block diagram of such a setup using an IBM PC or compatible as the test controller is shown in Fig. 6-8. (Such a system can also be assembled with an Apple Macintosh computer.) High-speed data transfer is not required in this case since the static linearity tests basically consist of a series of DC measurements. It is sufficient to read a byte of data from the A/D asynchronously at a low data rate suitable to the computer, while the DUT (device under test) repetitively samples the same input voltage.

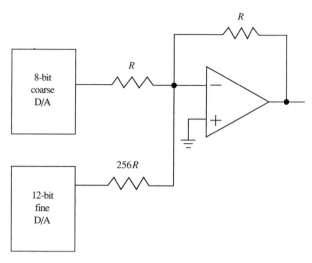

Figure 6-7 Coarse/fine D/A with 20-bit resolution.

The only additional hardware that is required to perform a PC-based digital static linearity test is a plug-in interface card with the addressing logic to communicate with the DUT. The test software consists of a loop where a D/A is periodically incremented (or decremented) to create a precise monotonic ramp input signal. One or several continuous sweeps of the A/D's range can be made without explicitly searching for code edges, as in the servo method. The test time will be reduced by eliminating the slow process of taking analog measurements with a DVM. Plug-in cards with 12-bit D/As are available to provide the input signal; for higher resolution the arrangement in Fig. 6-7 can be used. For this test a slow but very accurate D/A is used since there is plenty of time between samples for it to settle. The A/D can run at its highest sample rate, and as the output bytes are read back into the PC a running count is kept of the number of occurrences of each code. As an example, if a 12-bit D/A is used with an 8-bit A/D, in one sweep each acquired code should be roughly equivalent to 16 counts of the D/A (allowing for sufficient overlap on the end points). Storing the results requires a data array of $2^N - 2$ bytes, which will compose a linear histogram.

An example of a histogram plot for a 7-bit A/D is shown in Fig. 6-9. The number of occurrences is plotted on the y-axis versus each digital code on the x-axis. For an ideal A/D the distribution of occurrences of each code would be perfectly uniform, resulting in a horizontal straight line in the histogram plot. The ideal "width" of each code should be equal to the total number of samples acquired divided by the number of codes tested (i.e., for $n = 1$ to $2^N - 2$):

$$h(n)_{\text{IDEAL}} = \frac{N_{\text{tot}}}{2^N - 2}$$

To measure DNL from the linear histogram the following equation is used:

$$\text{DNL}(n) = \frac{h(n)_{\text{ACTUAL}} - h(n)_{\text{IDEAL}}}{h(n)_{\text{IDEAL}}} - 1$$

The integral nonlinearity can be calculated simply as the running total of the DNL calculations:

$$\text{INL}(1) = \text{DNL}(1)$$
$$\text{INL}(2) = \text{INL}(1) + \text{DNL}(2)$$
$$\text{INL}(3) = \text{INL}(2) + \text{DNL}(3) \quad , \ldots \text{etc.}$$

Examination of the definitions of INL and DNL will show that the relationship above is always true; but with a purely digital

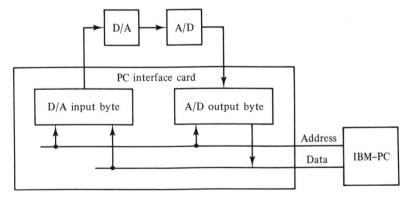

Figure 6-8   PC-based digital tester for static linearity measurement.

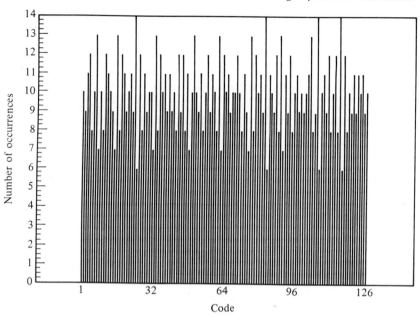

Figure 6-9   Linear histogram for 7-bit A/D.

method the accumulation of DNL measurement error in the running total, which limits its application in analog tests, is eliminated. Since the end points of the equivalent straight line for the INL calculation are at code 1 and code $2^N - 2$, INL($2^N - 2$) will always equal zero (i.e., DNL always sums to zero).

If measurements of offset and gain are required a DVM can be read just at the end points, or calibration of the D/A can be used as an analog measurement. To find the end-code transition points a method which is typically used is to repetitively collect sets of data at fixed increments of the D/A voltage. For measuring offset, a histogram for each input level is examined to find the point where an approximately equal population of code 0 and code 1 is achieved. If the A/D resolution is sufficient to avoid noisy transitions, the offset voltage will be obvious as the point where the number of samples of code 1 abruptly becomes greater than the number of samples of code 0. A similar procedure can be used at the full-scale end point, either between codes $2^N - 1$ and $2^N - 2$ or between $2^N - 1$ and the overflow signal transition.

If monotonicity or transition noise is a problem throughout the range of a particular A/D, the incoming data can be monitored to prevent such problems from being lost in the histogram. With a sufficient oversampling rate for each code, such as 16:1, a range of oscillating codes can be measured precisely and factored to give an equivalent percentage of an LSB. It may be desirable to reject devices which exhibit oscillatory behavior over a number of consecutive increments of the input voltage. Code transitions can be tested so that two consecutive samples never differ by more than ± 1. To reduce the effects of random noise in an otherwise well-behaved A/D, a larger histogram can be accumulated from multiple sweeps of the A/D's range.

Acquiring data from the A/D directly in digital form is useful in a variety of tests and becomes absolutely necessary for the dynamic tests which are described below. In some of these tests a sequence of output bytes from the A/D must be acquired synchronously at high speed so that the simple interface described previously is inadequate. For general purpose data acquisition, a logic analyzer is useful as a high-speed buffer between the A/D and the computer.

Logic analyzers, such as the DAS series of products from Tektronix, are available that can acquire data synchronously at speeds up to 2 GHz. By using a GPIB (general purpose interface bus) or IEEE-488 interface, binary data can be transferred to a computer at high speed for processing. Such instruments also provide the capability to send digital test vectors which can be used to control the input through a D/A. Some logic analyzers also feature a graphing capability, which is useful to preview the digitized data as it would appear when reconverted on an oscilloscope.

In order to do meaningful work with the data once it is transferred to the PC, an analytical software package is required. Appropriate packages range from complex equation solvers with notebooklike formats to integrated packages specifically for DSP analysis. Several software vendors provide a suite of signal analysis functions, along with graphics plotting capability and an interface for GPIB control. Many of the GPIB packages are intended to minimize the amount of programming that is required to perform data transfer and instrument control. Most major vendors of GPIB instruments and manufacturers of GPIB plug-in cards offer analytical and control software as well. Three of the most popular packages are Asystant (from Keithley Asyst, Rochester, New York), LabView (from National Instruments, Austin, Texas), and DADiSP (from DSP Development Corporation, Cambridge, Massachusetts). A block diagram of a complete digital tester for high-speed A/Ds is shown in Fig. 6-10.

Dynamic Linearity Tests

Sine-Wave Histograms: The Code-Density Test

Sine-wave histogram testing of an A/D converter involves collecting a large number of digitized samples over a period of time in order to build a model of the converter's response to a specific, well-defined input signal. By collecting a large number of samples from the A/D, the code-density model relies on an empirically developed probability distribution to predict the linearity of the converter. The accuracy of the model depends on a close matching of the actual A/D to the expected performance, since small numbers of large errors will be masked out. In fact, by relying on an assumption of randomness of the individual samples, this test method discards the position in time of an individual sample so that a determination of the accuracy of the quantization is not reported.

Because of the large number of samples that must be acquired, a histograming memory is often used to implement the code-density test directly in hardware. The acquired code will serve as the address of a byte in memory, where the count for that code (referred to as a "bin") is being accumulated. For each sample that is taken, a read/write cycle is executed and the content of the addressed byte is simply incremented by one. Rather than posing a requirement for a very deep memory at one byte per sample, in this way the limitation on the size of the dataset is the byte width of the N words of storage.

Since ultralinear ramp signals are very difficult to generate at high frequencies and measure for accuracy, sine waves are used

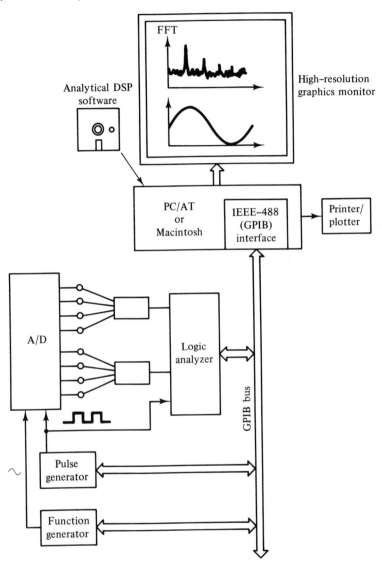

Figure 6-10 Complete PC-based digital test system for high-speed A/D converters.

in code-density testing. Ultrastable oscillators are generally available, and additional filtering can always be used to lower distortion. The purity of sine-wave signal sources can also be verified through the use of spectrum analyzers. Also, the mathematics of sine waves allows the model to be developed in a straightforward manner.

One change from ramp to sine-wave testing is that the probability of each code is no longer uniform, since the signal is nonlinear. For a sine wave with peak amplitude A, the probability of a voltage between any two points V_1 and V_2 is given by the following equation:

$$P(V_1 < V < V_2) = \frac{1}{\pi} \cdot [\sin^{-1}(V_1/A) - \sin^{-1}(V_2/A)]$$

The "cusp-shaped" probability distribution function (PDF) for a sine wave is shown in Fig. 6-11. In a unipolar A/D, with the code 1 transition offset to $V_{REF}/2^{N+1}$, the ideal threshold points of any code i between code 1 and 2^N-2 are

$$V_{lower} = i \cdot \frac{V_{REF}}{2^{N+1}}, \quad V_{upper} = (i + 2) \cdot \frac{V_{REF}}{2^{N+1}}$$

Thus the probability of any code i in the sine-wave code-density test of a unipolar A/D is

$$p_i = \frac{1}{\pi} \cdot \left[\sin^{-1}\left[\frac{(i + 2)}{2^{N+1}} \cdot \frac{V_{REF}}{A} \right] - \sin^{-1}\left[\frac{(i)}{2^{N+1}} \cdot \frac{V_{REF}}{A} \right] \right]$$

The signal amplitude A must be sufficient to insure clipping at both extremes of the A/D's range, and the signal must be offset so that the midpoint "zero-crossing" coincides with the midscale of the A/D. Even if these parameters are measured very accurately at the A/D input, the converter's gain and offset must be taken into account in the actual calculations of p_i. A PDF must be calculated for each A/D and for each set of test conditions.

Signal offset can be checked very easily by comparing the total number of samples, or cumulative histogram above and below the midpoint [ch (a,b)]. The signal offset can be adjusted until the

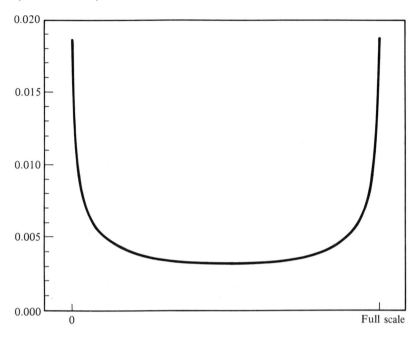

Figure 6-11   Probability distribution of a sine wave.

following condition is met:

$$ch(0, 2^{N-1} - 1) \approx ch(2^{N-1}, 2^N - 1)$$

The effective amplitude can be calculated by counting the number of samples that are clipped in BIN_0 and $BIN_{2^N - 1}$. An adjusted value can be estimated from the following equation:

$$A^* = \frac{V_{REF}}{\sin\left[\dfrac{N_{TOT} - h(0) - h(2^N - 1)}{N_{TOT}} \cdot \dfrac{\pi}{2}\right]}$$

The frequency of the input is chosen so that it is not harmonically related to the sampling clock, and it is sampled asynchronously to assure randomness. The real probability of occurrence of each code is then found from the count in each bin as $h(i)/N_{TOT}$. On the assumption that the calculated probability is a valid indication of code width, an estimate of differential nonlinearity is made from the following equation:

$$\text{DNL}_i = \frac{h(i)}{p_i \cdot N_{\text{TOT}}} - 1 \text{ LSB}$$

Sine-Wave Curve Fitting: Effective Bits

A curve fitting algorithm is often used to measure the "effective bits" of resolution of an A/D when digitizing a sine-wave input. Although similar in concept to the best-fit straight-line integral linearity measurement described in Chapter 3, the best-fit sine-wave calculation must determine all four parameters that describe a sine wave: amplitude, phase, frequency, and offset. Like the straight-line test, the fitted sine wave is used as a reference from which to calculate the error in the actual data.

If the frequency of the input signal that is digitized is known, the algorithm below can be used to calculate the best-fit sine wave. [The method described here is adapted from IEEE Std. 1057, "An Algorithm for Three Parameter (Known Frequency) Least Squared Fit to Sine-Wave Data," *IEEE Trial-Use Standard for Digitizing Waveform Recorders*, pp. 13–14.]

If a sequence of M digitized samples y_n are acquired with a sampling interval of t_n from an input sine wave of angular frequency ω, the first step is to compute the following nine sums:

$$\sum_{n=1}^{M} y_n \tag{1}$$

$$\sum_{n=1}^{M} \alpha_n \tag{2}$$

$$\sum_{n=1}^{M} \beta_n \tag{3}$$

$$\sum_{n=1}^{M} \alpha_n \cdot \beta_n \tag{4}$$

$$\sum_{n=1}^{M} \alpha_n^2 \tag{5}$$

$$\sum_{n=1}^{M} \beta_n^2 \qquad (6)$$

$$\sum_{n=1}^{M} y_n \cdot \alpha_n \qquad (7)$$

$$\sum_{n=1}^{M} y_n \cdot \beta_n \qquad (8)$$

$$\sum_{n=1}^{M} y_n^2 \qquad (9)$$

where $\alpha_n = \cos(\omega t_n)$, $\beta_n = \sin(\omega t_n)$. Using these sums, compute the following:

$$A = \frac{A_N}{A_D}$$

where

$$A_N = \frac{\sum_{n=1}^{M} y_n \cdot \alpha_n - \bar{y} \cdot \sum_{n=1}^{M} \alpha_n}{\sum_{n=1}^{M} \alpha_n \cdot \beta_n - \bar{\beta} \cdot \sum_{n=1}^{M} \alpha_n} - \frac{\sum_{n=1}^{M} y_n \cdot \beta_n - \bar{y} \cdot \sum_{n=1}^{M} \beta_n}{\sum_{n=1}^{M} \beta_n^2 - \bar{\beta} \cdot \sum_{n=1}^{M} \beta_n}$$

$$A_D = \frac{\sum_{n=1}^{M} \alpha_n^2 - \bar{\alpha} \cdot \sum_{n=1}^{M} \alpha_n}{\sum_{n=1}^{M} \alpha_n \cdot \beta_n - \bar{\beta} \cdot \sum_{n=1}^{M} \alpha_n} - \frac{\sum_{n=1}^{M} \alpha_n \cdot \beta_n - \bar{\alpha} \cdot \sum_{n=1}^{M} \beta_n}{\sum_{n=1}^{M} \beta_n^2 - \bar{\beta} \cdot \sum_{n=1}^{M} \beta_n}$$

The next step is to then compute:

$$B = \frac{B_N}{B_D}$$

where

$$B_N = \frac{\sum_{n=1}^{M} y_n \cdot \alpha_n - \bar{y} \cdot \sum_{n=1}^{M} \alpha_n}{\sum_{n=1}^{M} \alpha_n^2 - \bar{\alpha} \cdot \sum_{n=1}^{M} \alpha_n} - \frac{\sum_{n=1}^{M} y_n \cdot \beta_n - \bar{y} \cdot \sum_{n=1}^{M} \beta_n}{\sum_{n=1}^{M} \alpha_n \cdot \beta_n - \bar{\alpha} \cdot \sum_{n=1}^{M} \beta_n}$$

$$B_D = \frac{\displaystyle\sum_{n=1}^{M}\alpha_n \cdot \beta_n - \bar{\beta} \cdot \sum_{n=1}^{M}\alpha_n}{\displaystyle\sum_{n=1}^{M}\alpha_n^2 - \bar{\alpha} \cdot \sum_{n=1}^{M}\alpha_n} - \frac{\displaystyle\sum_{n=1}^{M}\beta_n^2 - \bar{\beta} \cdot \sum_{n=1}^{M}\beta_n}{\displaystyle\sum_{n=1}^{M}\alpha_n \cdot \beta_n - \bar{\alpha} \cdot \sum_{n=1}^{M}\beta_n}$$

then compute

$$C = \bar{y} - A \cdot \bar{\alpha} - B \cdot \bar{\beta}$$

where the mean values are calculated as

$$\bar{y} = \frac{1}{M} \cdot \sum_{n=1}^{M} y_n$$

$$\bar{\alpha} = \frac{1}{M} \cdot \sum_{n=1}^{M} \alpha_n$$

$$\bar{\beta} = \frac{1}{M} \cdot \sum_{n=1}^{M} \beta_n$$

From the previous calculations a fitted function is then produced as

$$y'_n = A \cos(\omega t_n) + B \sin(\omega t_n) + C$$

The RMS error of this fitted function relative to the acquired data can be calculated as follows:

$$v_{e,RMS} = \sqrt{e/M}$$

where

$$e = \sum_{n=1}^{M} y_n^2 + A^2 \cdot \sum_{n=1}^{M} \alpha_n^2 + B^2 \cdot \sum_{n=1}^{M} \beta_n^2 + M \cdot C^2$$
$$- 2A \sum_{n=1}^{M} \alpha_n \cdot y_n \quad - 2B \sum_{n=1}^{M} \beta_n \cdot y_n$$
$$- 2C \sum_{n=1}^{M} y_n \quad + 2AB \sum_{n=1}^{M} \alpha_n \cdot \beta_n$$
$$+ 2AC \sum_{n=1}^{M} \alpha_n \quad + 2BC \sum_{n=1}^{M} \beta_n$$

To convert the amplitude and the phase to the form

$$y_n = A_{\cos}\cos(\omega t_n + \theta) + C$$

use

$$A_{\cos} = (A^2 + B^2)^{\frac{1}{2}}$$

$$\theta = \tan^{-1}\left(-\frac{B}{A}\right)$$

A more complete derivation of this algorithm is contained in Appendix A of IEEE Std. 1057.

In an ideal A/D the quantization error is uniformly distributed across all codes. Regardless of the signal type, the RMS value of this error was found in Chapter 3 to be

$$v_{e,\text{RMS}} = q/\sqrt{12}$$

where by definition q is 1 LSB or $V_{\text{REF}}/2^N$. To find the effective number of bits of resolution in an equivalent ideal A/D, the equation above becomes

$$v_{e,\text{RMS}} = \frac{V_{\text{REF}}}{\sqrt{12} \cdot 2^{N_{\text{eff}}}}$$

so that

$$N_{\text{eff}} = \log_2\left[\frac{V_{\text{REF}}}{\sqrt{12} \cdot v_{e,\text{RMS}}}\right]$$

The RMS error, which was calculated from the best-fit sine wave in previous equations, can be used in this equation. An effective bits measurement is also sometimes derived from the measured versus ideal signal-to-noise ratio in the digitized data. This measurement requires a spectral analysis which is performed in the frequency domain using the transform methods that are discussed later in this chapter.

Envelope and Beat Frequency Tests

The envelope test can be used as a qualitative measure of large-signal bandwidth (sometimes referred to as full-power bandwidth) and slewing ability in flash A/D converters. In this test a full-scale sine wave at a frequency slightly higher than the Nyquist frequency, or one-half the sampling rate, is used as the input signal to the A/D. This relationship between the sampling clock and input signal produces two digitized outputs for each period of the input, offset in phase by approximately 180°, as shown in Fig. 6-12.

By using a reconversion D/A, or plotting the digitized outputs collected from a logic analyzer, the "envelope" will appear as two low-frequency sine waves, which are amplitude modulating the input frequency, 180° out of phase. Though not truly AM signals, the output waveforms will be the difference between the input frequency f_{in} and the Nyquist rate $f_{CLK}/2$. As a measure of large-signal response and overload recovery, this test presents a worst-case situation to the A/D when the peaks of the input are being sampled. At these points the A/D must slew to opposite extremes of its reference range on consecutive samples. Nonsymmetrical slewing ability of the A/D is indicated by asymmetry on either side of the peaks of the envelope.

At the zero crossing of the envelope (which should coincide with midscale for the A/D), noise effects, aperture uncertainty, slewing limitations, and sparkle codes may become evident. Sampling at this point exposes the comparators to the fastest slewing portion of the input signal. Nonsymmetry in the comparator's response will be seen by comparing alternate zero crossings. As the input crosses the threshold of the MSB transition, aperture errors or poor thermometer decoding between adjacent comparators may cause an output glitch or "sparkle code."

Although the output of the envelope test is usually not measured directly for linearity, a test for missing codes under dynamic conditions can be done by properly choosing the input frequency. For this test the envelope frequency $\Delta f = f_{in} - f_{CLK}/2$ should be low enough to provide several samples per code at the maximum slewing point:

$$\frac{dV_{in}}{dt} = \pi \cdot \Delta f \cdot V_{REF} = \frac{V_{REF}}{x \cdot 2^N} \cdot \frac{1}{2 \cdot t_{CLK}}$$

$$\Delta f = \frac{1}{x \cdot 2^N \cdot \pi} \cdot \frac{f_{CLK}}{2}$$

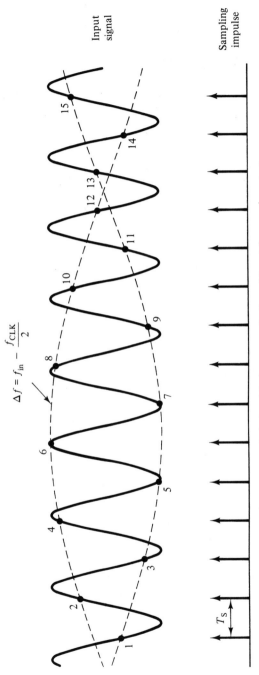

Figure 6-12 Relationship between clock and input for envelope test.

Using these calculations, the envelope frequency Δf will be a fraction of the sampling clock frequency sufficient to produce x samples per code at the zero crossing.

If an A/D performs satisfactorily in an envelope test, the concept can be extended to the beat-frequency test as a further measure of input signal bandwidth. The beat frequency is generated by the aliasing of a signal offset slightly in frequency from the sampling rate, as shown in Fig. 6-13. For this test a single tone will be observed at the difference frequency between the input and the sampling clock. The equation above can be used to assure that all codes are sampled by changing the $f_{CLK}/2$ term to f_{CLK} in the calculation of Δf.

If a reconversion D/A is used to observe the A/D's output on an oscilloscope, attenuation due to the internal bandwidth limitations of the A/D can be directly measured in real time during the beat and envelope tests. By calibrating the oscilloscope graticule against the output of a full-scale low-frequency signal, loss of dynamic range with an aliased signal at the same frequency will be obvious. Of course, the same measurement can also be made by examining the extent of the digital codes that are acquired from a logic analyzer.

For A/Ds with extended signal bandwidth capability, the envelope and beat-frequency tests can be repeated at higher multiples of the Nyquist rate. As shown in Fig. 6-14, odd integer multiples produce the envelope of two tones at 180° to each other, while even multiples produce single tones.

FFT Testing

The FFT Algorithm

The well-known Fourier transform is used to convert a continuous time-domain signal into a representation of its components in the frequency domain. The frequency components consist of complex exponential functions, which can also be expressed as combinations of sine and cosine functions. The transform output yields an amplitude and phase spectrum describing all the frequency components of the original signal. To obtain such infor-

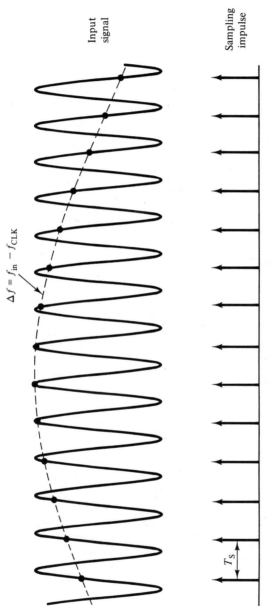

Figure 6-13 Relationship between clock and input for beat-frequency test.

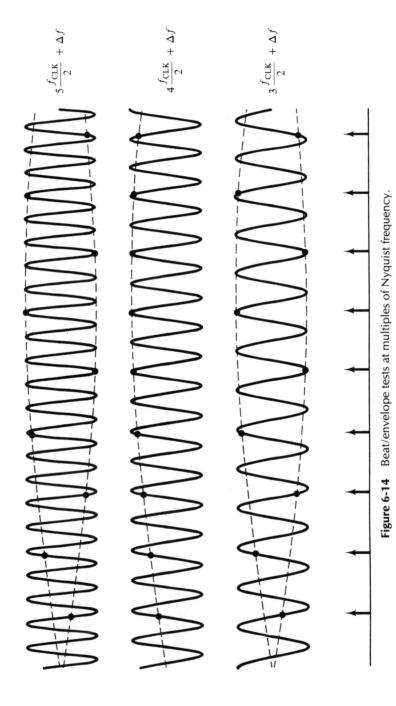

$5\dfrac{f_{\text{CLK}}}{2} + \Delta f$

$4\dfrac{f_{\text{CLK}}}{2} + \Delta f$

$3\dfrac{f_{\text{CLK}}}{2} + \Delta f$

Figure 6-14 Beat/envelope tests at multiples of Nyquist frequency.

mation from the quantized output of an A/D converter, which is a discrete-time signal, transformation based on the discrete Fourier transform (DFT) is performed. If a total of N samples of a periodic signal are acquired at fixed time intervals T, then the terms of the DFT are calculated from the following equation:

$$X_k = \sum_{n=0}^{N-1} x(nT) \cdot e^{-j2\pi nk/N}$$

The DFT converts N samples of a signal in the time domain into a complex function in the frequency domain, which will be periodic with a frequency of $\omega_s = 2\pi/T$. A total of N discrete X_k terms result over the period of ω_s. Calculation of the magnitude of this function, which forms the amplitude spectrum, shows that it is symmetrical such that

$$|X_k| = |X_{N-k}|$$

An offset of $f_s/2$ applied to the transformed spectrum produces $N/2$ unique frequency terms plus the DC term at a frequency of zero. The negative frequencies provide redundant information and can be discarded. Each discrete frequency term will be separated by f_s/N in the output spectrum, over the bandwidth of zero, to $f_s/2$, which corresponds with the limitations of the Nyquist sampling theory.

The equation above for calculating the DFT results in N complex multiply operations and $N - 1$ complex additions for each of the X_k terms. For the entire transform this would result in N^2 multiplications and $N \cdot (N - 1)$ additions. This large number of complex operations would be too slow in most applications, so fast Fourier transform (FFT) algorithms have been developed to reduce the number of calculations that are required. Most of these algorithms exploit the symmetry properties of the DFT to eliminate redundant calculations. The popular "radix-2" approach requires $\log_2(N)$ cycles of calculations, with each cycle containing $N/2$ complex additions and $N/2$ complex multiplications. This dramatically reduces the total number of calculations from $2 \cdot N^2 - N$ to $N \cdot \log_2(N)$. For example, a 1K-point FFT requires only 10,240 complex operations, as compared to the

DFT, which would require 2,096,128 operations. To utilize such algorithms, the number of samples that are collected must equal an integer power of two. An example of an amplitude spectrum generated from an FFT algorithm is shown in Fig. 6-21.

Windowing Functions

The DFT is based on an assumption of periodic signals (i.e., that the sequence of N samples repeats to infinity). Since in reality the number of samples is finite, the resulting spectral analysis may be corrupted by this erroneous assumption. This will occur because of discontinuities in the end points of the data sequence, as illustrated in Fig. 6-15. For a periodic waveform, such as the sine wave in the illustration, repetition of the data sequence may show that the first and last samples are not consecutive points on the waveform. This will be the case whenever the length of the data sequence does not coincide with a whole number of periods of the

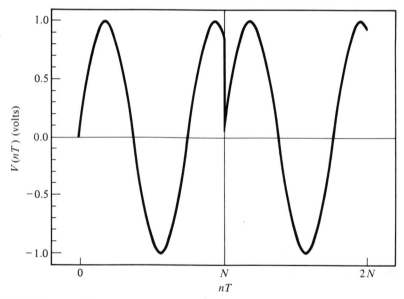

Figure 6-15   Discontinuities result from sampling less than a whole number of periods of the input.

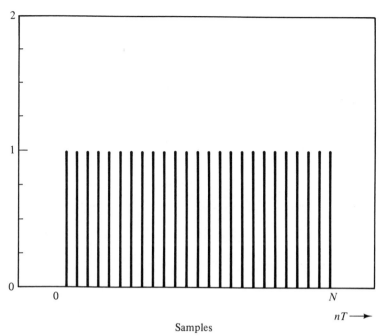

Samples

Figure 6-16   Sampling effect of rectangular window function.

input signal. The effect of transforming these discontinuities caused by the end points is known as spectral leakage.

Another way of understanding the end point effect in the FFT is to observe that the sampling process in effect multiplies the input by a rectangular window of unity height impulses, as shown in Fig. 6-16. This discrete-time function has a value in the time domain represented by the following equation:

$$W_{\text{rect}}(nT) = 1; \qquad 0 \leq nT \leq N \cdot T$$
$$W_{\text{rect}}(nT) = 0; \qquad nT < 0, nT > N \cdot T$$

This multiplication in the time domain is equivalent to convolution in the frequency domain, with the result that the transformed spectrum of each frequency component will be "spread" by the transformation of the window function. The amplitude spectrum of the rectangular window is the $\sin(x)/x$, or sinc function, which is shown in Fig. 6-17a. The side lobes of the sinc

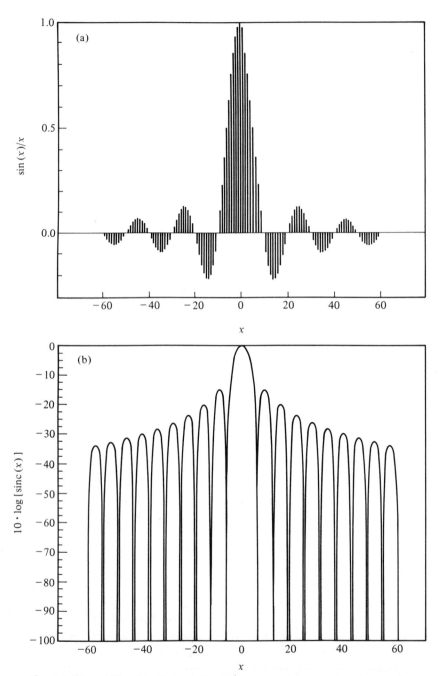

Figure 6-17 (a) Sinc function amplitude spectrum of rectangular window. (b) Log magnitude of sinc function. (c) Skirting effect on ideal sine wave. (*Figure continues.*)

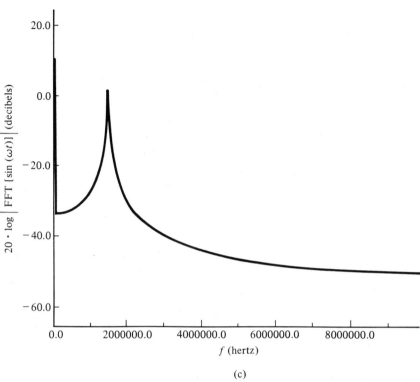

(c)

Figure 6-17 (continued)

function cause spectral leakage of a discrete frequency to adjacent frequency terms. On a log plot with the y-axis in decibels, the rectangular window exhibits a "skirting" effect as shown in Fig. 6-17b. The skirting effect is clearly evident in Fig. 6-17c, where the amplitude spectrum of an ideal sine wave, which contains no energy outside of its fundamental frequency, is spread over the entire Nyquist bandwidth.

There are three alternatives to choose from in order to obtain meaningful results from the FFT. In a test setup where the choice of input frequency is arbitrary, the first two options force the input to equal a whole number of periods over the sampling interval. In other words, the following relationships are observed:

$$M \cdot T_{in} = N \cdot T_s$$

or

$$\frac{f_{in}}{M} = \frac{f_s}{N}$$

The value of M must be an integer and, as required for the FFT, N is an integer power of 2. To maximize the number of codes that are covered and eliminate redundancy in the data, M should be odd and prime. One way of generating this relationship is to build a divider and phase-locked loop (PLL) frequency synthesizer synchronized to a precision reference, as shown in Fig. 6-18. This method generates coherent sampling but is limited in flexibility by the choice of divider ratios.

Coherent sampling is not required if a synthesized function generator is available to supply the input signal, such as the HP3325A. Such a generator provides resolution down to a fraction of a hertz, which is more than sufficient to set the signal frequency to eliminate skirting effects. In most cases, while the signal is swept over various frequencies to cover the input bandwidth, the clock frequency is a nice round number. If desired, both the clock and signal can still be synchronized to the same reference oscillator. A simple calculation is all that is necessary to choose the signal frequency by finding an appropriate value for M. The following example will illustrate this method:

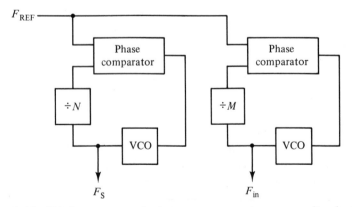

Figure 6-18   PLL frequency synthesizer to generate coherent sampling for FFT.

Step 1. *Using a first-pass choice for* f_{in} *(1 MHz), calculate the value for* M *for a given number of samples* N *(1024 or* 2^{10}*) and sampling frequency* f_s *(20 MHz):*

$$M = \frac{f_{in}}{f_s/N} = \frac{10^6 \cdot 1024}{20 \cdot 10^6} = 51.2$$

Step 2. *Adjust the value of* M *to the nearest odd integer so that it coincides with one of the discrete frequencies of the FFT and recalculate* f_{in}:

$$M' = 51$$
$$f_{in} = \frac{51 \cdot 20 \cdot 10^6}{1024} = 996{,}093.75 \text{ Hz}$$

The third option to reduce spectral leakage, which is particularly appropriate when the input signal is unknown beforehand, is to apply a different window function to the data which contains less energy in the side lobes. Some of the popular functions for this purpose are the Hann, Hamming, and Blackman–Harris windows. The window functions will gradually smooth the end points of the data to remove discontinuities at the window edges. The shape of the Hamming window function is shown in Fig. 6-19.

The Hann and Hamming windows are characterized by the following equation:

$$W(n) = \alpha + (1 - \alpha) \cdot \cos\left[\frac{2\pi \cdot n}{N}\right]$$

For the Hann window the coefficient α is 0.5, while for the Hamming window it is 0.54.

These functions will all provide much greater attenuation to the side lobes at the expense of a somewhat wider main lobe response. This effect is apparent by comparing the log magnitude of the Hamming function in Fig. 6-20 to the sinc function. When analyzing an A/D's response to sine waves, the window functions will reduce the fundamental and harmonics from a single line, as in the ideal case, spreading some of the energy to adjacent frequencies. The particular characteristic of the function being

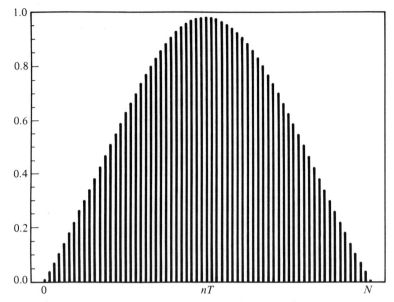

Figure 6-19   Hamming window function.

used must be taken into account when calculating signal-to-noise ratios from the amplitude spectrum. For a thorough comparison of the various window functions, see the reference by Harris in the bibliography.

Interpreting the FFT Amplitude Spectrum

Figure 6-21 illustrates a typical FFT analysis of an A/D converter. The sampling frequency is set at 20 MHz, and the data record size will be 1024 samples. An input frequency of approximately 1.5 MHz will be digitized:

$$M = (1.5 \cdot 10^6 \cdot 1024)/(20 \cdot 10^6) = 76.8$$

Using $M' = 77$, the equation is

$$f_{in} = 77 \cdot 20 \cdot 10^6/1024 = 1,503,906.25 \text{ Hz}$$

After transformation, the magnitude spectrum contains 512 discrete frequency components between zero and the Nyquist limit of 10 MHz. The spacing between frequencies is 20 MHz/ 1024 = 19,531.25 Hz. The process of adjusting the input frequency to a whole number of periods in the sampling window is equivalent in the frequency domain to selecting the spectral line at which the fundamental component will be placed. Through the calculation above we have simply forced the input frequency to exactly match one of the discrete frequencies which the FFT is capable of producing. In this case the fundamental input frequency is at line 77 of the output spectrum.

Harmonics of the input signal will show up at integral multiples of the fundamental. In Fig. 6-21, a second harmonic is evident at line 154 (3,007,812.5 Hz), and a sixth harmonic is seen at line 462 (9,023,437.5 Hz). What happens to the higher order harmonics? Figure 6-21 presents a graphic demonstration of aliasing which results from the Nyquist sampling theory.

Although only part of the spectrum is shown in the FFT plot, the sampling theory states that the spectrum of a signal will be

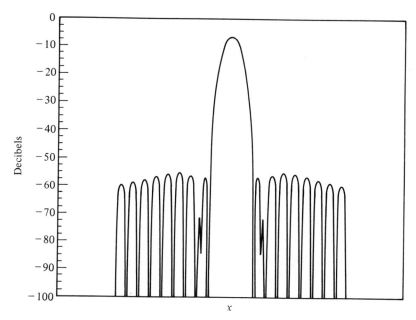

Figure 6-20 Log magnitude of Hamming window function.

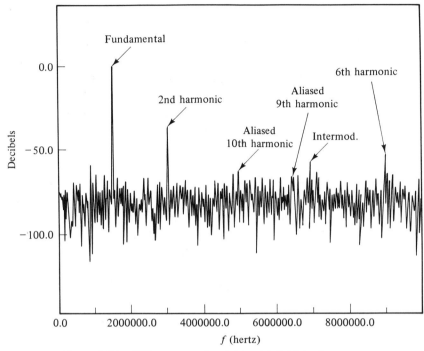

Figure 6-21   FFT spectrum for digitized 1.5-MHz sine wave.

reproduced at 0, $\pm f_s$, $\pm 2f_s$, etc. Figure 6-22 shows the position of harmonic components in the signal from the earlier example. The solid lines represent the input components. With a 1.5-MHz input and a sampling frequency of 20 MHz, only the first six harmonics are in the Nyquist bandwidth. If the signal contains components beyond Nyquist, they are folded back, or aliased. This is represented by the dashed lines which form a mirror-image spectrum centered at f_s. Notice that the out-of-band harmonics show up at positions which are mirrored about the Nyquist frequency.

Returning to Fig. 6-21, a ninth harmonic would be produced at a position of 9·77, or line 693. This frequency is mirrored at $1024 - 693 = 331$. The aliased frequency is 6,464,843.75 Hz. To calculate total harmonic distortion (THD), this method is used to locate all of the significant harmonic frequencies.

Calculation of signal-to-noise ratio, or signal-to-noise + distortion, can be performed by determining the RMS value of the spectral components minus the fundamental. If a windowing function causes spreading of the fundamental, this effect should also be taken into account. An alternative calculation of effective bits is sometimes made on the basis of SNR, as was discussed in Chapter 3. This equation is repeated below:

$$\text{effective bits} = \frac{\text{measured SNR} - 1.76}{6.02}$$

Another specification based on the FFT spectrum, which is often used to represent the effective resolution of an A/D, is the spurious-free dynamic range. This is simply the ratio in decibels between the magnitude of the fundamental component and the magnitude of the largest harmonic or intermodulation product. The spurious-free dynamic range can also be defined as the usable dynamic range, since it describes the minimum amplitude of input signals which can be reliably identified without the possibility of distortion caused by A/D errors.

Aperture Errors—Locked Histograms

The locked histogram test can be used to estimate the aperture jitter and uncertainty in a flash A/D. To perform this test the same frequency is used for both the sampling clock and input sine

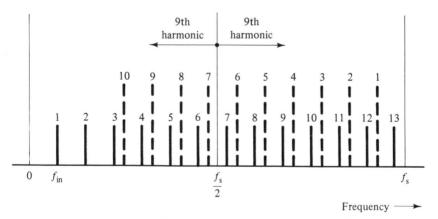

Figure 6-22 Folding of spectrum due to aliasing in FFT.

wave. A full-scale or greater analog input is synchronized to the sampling clock such that the same point on the input is repetitively sampled. The stability of the synchronizing signal must be greater than the error which is being measured. One setup which can be used to generate this test is shown in Fig. 6-23.

The effects of aperture jitter will become more evident as the signal slew rate increases. With a given clock and input frequency, a delay line is used to shift the sampling point to near midscale, where the rate of change of the sine wave is greatest. This can be done to sufficient precision in many applications by simply adjusting the length of the coaxial cable in either the clock or input line. Based on the speed-of-light propagation, the following delay calculation can be used:

$$t_D = \frac{1}{3 \cdot 10^8 \text{ m/sec}}$$
$$t_D = 3.33 \text{ nsec/meter}$$

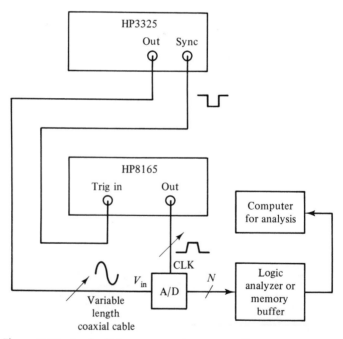

Figure 6-23   Locked histogram test for aperture jitter ($F_{in} = F_{CLK}$).

A high-speed buffer memory is used to collect the data, as discussed earlier. If a logic analyzer is used, the state table display makes it easier to tune the delay line until output codes near midscale are acquired. The data can then be transferred to a computer, where the histogram of the code distribution can be plotted. If the input voltage is close to a code transition point, even the best A/D will produce an output which alternates between two adjacent codes. However, for many well-behaved flash A/Ds, it may actually be difficult to get a histogram wide enough to allow meaningful statistics to be calculated on the distribution when the A/D is operated within its rated speed of conversion. In such cases, either the clock/input signal frequency or the amplitude of the signal must be increased to create a higher slew rate. A typical distribution that can be used for this calculation is shown in Fig. 6-24.

In order to estimate the A/D's aperture error, the standard deviation σ of the histogram is typically used. This is equivalent to the RMS aperture error. The following equations illustrate the calculation:

$$\text{Slew rate} = V_{\text{AMP}} \cdot 2\pi \cdot f_{\text{in}}$$
$$\text{Aperture error} = T_{\text{a}} = \frac{\sigma \cdot V_{\text{LSB}}}{\text{Slew rate}}$$

Noise-Power Ratio Test

The noise-power ratio (NPR) test has been used primarily to measure the performance of frequency division multiplexed (FDM) communication systems. In FDM communications, a spectrum of some limited bandwidth is divided into smaller frequency "slots," or channels, so that it can be shared amongst a number of simultaneous transmitters and receivers. In the typical example of a voice communication system, the individual channels are 4-KHz wide. The NPR test is used to measure the level of interference or noise from the signals present in the broader spectrum into a single channel. A standard to measure this level has been developed based on observations that as a large number (on the order of 100) of simultaneous channels are in use, the broad-band signal resembles noise with a Gaussian distribution.

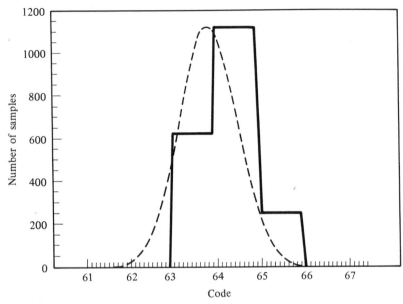

Figure 6-24   Results of locked histogram test.

Specialized equipment for this test (from manufacturers such as Marconi) models these conditions.

Although the information content in a single FDM channel is low in frequency, the high base-band frequencies require the use of high-speed A/D converters in DSP-based systems. A block diagram of a noise-power ratio test for a system using an A/D converter is shown in Fig. 6-25. The NPR test is one of the specialized measurements which is still best performed using analog techniques. The noise generator produces a band-limited Gaussian signal of the appropriate bandwidth and power. In the first step of the NPR test this signal is fed directly to the A/D converter input, where it is quantized and then reconverted to analog by a high-speed D/A. The D/A must itself be sufficiently low in noise so as to not corrupt the results. The noise receiver contains selectable band-pass filters to tune into the desired frequency slot. The receiver can then be calibrated so that 0 dB correspond to the power of the received signal.

In the second step, a selectable notch filter, which is matched to the frequency of the receiver channel, is switched in at the transmitter preceding the A/D converter. Ideally, this removes all signal content in the tuned channel, leaving only the remainder of the broad-band spectrum outside this channel, which is filtered out at the receiver. The power of any residual noise in the tuned channel is measured, and the ratio to the power of the signal measured in the first step then determines the NPR. The test is repeated at a number of channels throughout the range of the FDM spectrum. This test combines the effects of all sources of dynamic nonlinearity, intermodulation, and noise in the A/D into a single measurement.

The rigor of the NPR test can be compared to SNR or effective-bits tests, where filters are usually applied to limit the input signal of each measurement to a single frequency. As in these other tests, there is a theoretical limit on the NPR level which can be obtained for an A/D of a given resolution. In contrast to other tests, the NPR test is not done with a full-scale input signal. Analysis of the A/D conversion process on signals with a Gauss-ian probability distribution has shown that there is an optimum relationship between the RMS value of the input signal and the full-scale range of the converter which achieves a maximum NPR. (See the reference by Gray and Zeoli for more details).

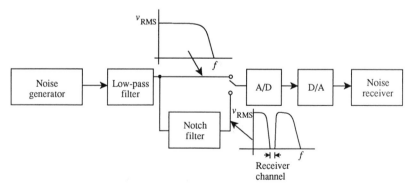

Figure 6-25   Measurement setup for testing noise-power ratio of an A/D con-verter.

Figure 6-26 is representative of a signal with a Gaussian distribution. To represent the relationship of the RMS amplitude of the noise signal to the peak full-scale limit of the A/D, a parameter referred to as the loading factor k is used:

$$k = \frac{V_{FS}}{\sigma}$$

where the A/D range is $\pm V_{FS}$ and σ is the standard deviation, or RMS amplitude, of the noise level. For the example in the illustration, σ was set to one. By rearranging terms so that V_{FS} is equated to $k\sigma$, Fig. 6-26b shows the problem that occurs with low loading factors. If k is too small, conditions will exist where there is a significant probability of the input signal level exceeding the range of the A/D. The result will be clipping in the digitized output, commonly referred to as saturation noise, causing a severe degradation in NPR.

For signals below the full-scale limit, the quantization noise limits NPR, with an inverse linear relationship between noise and signal level. However, for a value of k that is too large, an effective loss of resolution occurs such that a large percentage of signals are quantized over only a portion of the A/D's range. If saturation occurs, noise increases with an exponential relation-

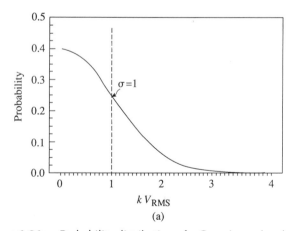

Figure 6-26a   Probability distribution of a Gaussian noise signal.

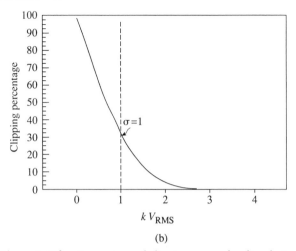

Figure 6-26b   Percentage of clipping-versus-loading factor.

ship versus signal level. The optimum value of k depends on the resolution of the A/D, ranging from 4 to 5 for 8-bit to 12-bit converters.

The maximum theoretical limit on NPR is determined the same way as SNR, which was derived in Chapter 3. In this case, the maximum RMS input σ_{max} is V_{FS}/k, which is equivalent to $2^{N-1} \cdot q/k$:

$$\text{NPR}_{max} = \frac{2^{N-1} \cdot q/k}{q/\sqrt{12}} = \frac{2^{N} \cdot \sqrt{3}}{k}$$

To convert NPR to decibels, the following equation results:

$$\begin{aligned}\text{NPR}_{max} &= 6.02 \cdot N + 20 \cdot \log(\sqrt{3}/k) \\ &= 6.02 \cdot N + 4.77 - 20 \cdot \log(k)\end{aligned}$$

To accommodate the Gaussian probability distribution of the input signal, the loss in dynamic range, or peak noise ratio, is $-20 \cdot \log(k)$. When conditions are optimized for a specified RMS input power, the loading factor then establishes the saturation point of the communication channel. The usable dynamic range of a FDM system will be set below this limit rather than the full-scale voltage of the A/D.

Differential Phase and Gain Measurements

For digital video systems, the measurement of differential gain (DG) and differential phase (DP) errors is still usually done with traditional analog instruments designed specifically for the purpose. As shown in Fig. 6-27, this equipment consists of a composite video signal generator and a specialized oscilloscope, which is referred to as a vectorscope. The signal generator is capable of producing a variety of TV test signals, such as color bars, crosshatch patterns, and gray scales. For the DG and DP measurements, a signal like the 10-step, 20-IRE modulated staircase illustrated in Fig. 3-12 may be used. Instead of discrete luminance steps, the modulated chrominance signal may instead be superimposed on a continuous ramp.

In the NTSC video standard, the frequency of the color subcarrier is 3.579545 MHz. The sampling rate of the A/D converter is recommended to be at least three to four times the subcarrier frequency. A clock rate should be chosen that is not harmonically related to the input signal repetition rate or subcarrier frequency to insure randomness and coverage of the complete A/D converter range. The output of the D/A converter should be filtered to avoid glitches in the digitized output. The NTSC video bandwidth is limited to 4.2 MHz.

The DG and DP measurements are very inexact because they require a visual interpretation of the error magnitude. The vectorscope graticule resolution is only 1% for DG and 1° for DP, although it is common to see specifications to one-tenth of these values. A typical vectorscope display is shown in Fig. 6-28. Initial calibration is produced by connecting the source video signal straight through to the vectorscope. With a sampling rate unlocked from the input signal, the error waveform will be a composite of random errors from repetitive sampling of a complete horizontal video scan line, which are visually integrated by the operator. As shown in the Fig. 6-28, this error appears as a fuzzy

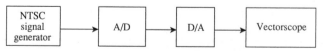

Figure 6-27   Test setup for differential-gain and differential-phase measurements.

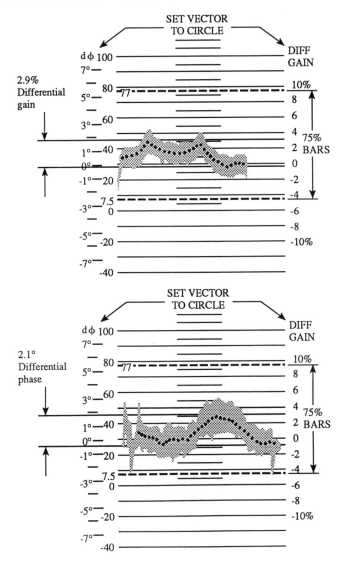

Figure 6-28 Vectorscope measurement of differential phase and differential gain. (Adapted with permission from IEEE Standard 746-1984, IEEE standard for performance measurements of A/D and D/A converters for PCM television video circuits.)

bar on the vectorscope display. The peak-to-peak amplitude will depend on the bits of resolution in the A/D converter. The typical procedure is to attempt to visualize the center line of the error waveform in order to measure the deviation from a perfect straight line. This is represented by the dotted line in the figure.

The bottom of the center line is set to 0% for the differential gain measurement. The positive peak deviation of the center line gives the amount of error, not the peak to peak of the overall waveform. (One-half of the peak to peak of the complete waveform could also be used.) A similar procedure is used for measurement of differential phase, which, although it uses the same input signal, is a separate setting on the vectorscope.

Bibliography

This bibliography represents an extensive collection of background material for readers who are interested in researching the topics discussed in this book in greater detail. These references have all been used by the author both in practice and in the preparation of this text. For those with an appetite for the most in-depth descriptions of high-speed A/D design, testing, and signal processing theory, the various IEEE journals and conference proceedings are highly recommended. For more application-oriented engineers, the trade magazines together with literature from manufacturers provide a steady flow of practical information.

The references have been arranged by chapter and subject in chronological order to provide a historical perspective of the evolving technology. While innovation continues to redefine the state of the art, many of the older papers still provide a wealth of knowledge on the fundamentals, which do not become obsolete.

Chapter 1

Dual-Slope A/Ds

Musa, F. H. and Huntington, R. C. (1976). A CMOS monolithic 3½-digit A/D converter. *IEEE International Solid-State Circuits Conference Digest of Technical Papers* **XIX:** 144–145.

Gordon, B. M. (1978). Linear electronic analog/digital conversion architectures, their origins, parameters limitations, and applications. *IEEE Trans. Circuits and Systems* **CAS-25** (7): 391–418.

Zuch, E. L. (ed.) (1979). "Data Acquisition and Conversion Handbook" GE Datel.

Renschler, E. "Analog-to-digital conversion techniques," Motorola Semiconductor Products Inc., application note AN-471.

Bloom, S. and Wayne, S. (1984). High-resolution A/D applications and methods proliferate. *Electronic Products* (December 12) pp. 65–72.

Delta-Sigma A/Ds

Rose, C. D. (1986). A new way to cut the cost of A-to-D converters. *Electronics* (March 31) pp. 42–44.

Koch, R., Heise, B., Eckbauer, F., Englehardt, E., Fisher, J. A., and Parzefall, F. (1986) A 12-bit sigma-delta analog-to-digital converter with a 15-MHz clock rate. *IEEE J. Solid-State Circuits* **SC-21** (6) 1003–1010.

"CMOS data acquisition circuits applications seminar." Crystal Semiconductor Corp., Austin, Texas, April 1987.

Matsuya, Y., Uchimura, K., Iwata, A., Kobayashi, T., Ishikawa, M., and Yoshitome, T. (1987). 16-bit oversampling A/D conversion technology using triple-integration noise shaping. *IEEE J. Solid-State Circuits* **SC-22** 921–929.

Riezenman, M. (1988). Special report: high-resolution arrives for ADC chips and hybrids. *Electronics* (January 7) 133–138.

Goodenough, F. (1988). Grab distributed sensor data with 16-bit delta-sigma ADCs. *Electronic Design* **36** (April 14).

Boser, B. E. and Wooley, B. A. (1988). The design of sigma-delta modulation analog-to-digital converters. *IEEE J. Solid-State Circuits* **23** (6) 1298–1308.

Rebeschini, M., van Bavel, N. R., Rakers, P., Greene, R., Caldwell, J., and Haug, J. R. (1990). A 16-b 160-kHz CMOS A/D converter using sigma-delta modulation. *IEEE J. Solid-State Circuits* **25** (2) 431–440.

Successive Approximation A/Ds

Gordon, B. M. *See* Dual-Slope A/Ds.

Henry, T. W. "High-speed digital-to-analog and analog-to-digital techniques." Motorola Semiconductor Products Inc., application note AN-702.

Bloom, S. and Wayne, S. *See* Dual-Slope A/Ds.

Dooley, D. J. (ed.) (1980). "Data Conversion Integrated Circuits" IEEE Press The Institute of Electrical and Electronics Engineers, Inc., New York.

Boyacigiller, Z. and Sockolov, S. (1982). Increase analog-system accuracy with a 14-bit monolithic ADC. *EDN* **27** (August 18) 137–144.

Bacrania, K. (1986). Digital error correction to increase speed of successive approximation. *IEEE ISSCC Dig. Tech. Papers* **XXIX** 140–141.

Sidman, S. and Harris, S. (1987). Hardware methods improve l-chip A/D converters. *EDN* **32** (February 5) 139–156.

Renschler, E. *See* Dual-Slope A/Ds.

Pipelined A/Ds

Lewis, S. H. and Gray, P. R. (1987). A pipelined 5-Msample/s 9-bit analog-to-digital converter. *IEEE J. Solid-State Circuits* **SC-22** (6) 954–961.

Goodenough, F. (1988). Analog ICs. *Electronic Design* **36** (February 18) 63–68.

Sutarja, S. and Gray, P. R. (1988). A pipelined 13-bit, 250-ks/s, 5-V analog-to-digital converter. *IEEE J. Solid-State Circuits* **23** (6) 1316–1323.

Song, B.-S., Tompsett, M. F., and Lakshmikumar, K. R. (1988). A 12-bit l-Msample/s capacitor error-averaging pipelined A/D converter. *IEEE J. Solid-State Circuits* **23** (6) 1324–1333.

Algorithmic A/Ds

McCharles, R. H. *et al.* (1977). An algorithmic analog-to-digital converter. *IEEE International Solid-State Circuits Conf. Digest of Tech. Papers* **XX** 96–97.

Li, P. W. (1984). "Ratio-Independent Algorithmic Analog-to-Digital Conversion Techniques," PhD thesis, UC Berkeley Memorandum no. UCB/ERL M84/66 (August 20).

Li, P. W., Chin, M. J., Gray, P. R., and Castello, R. (1984). A ratio-independent algorithmic analog-to-digital conversion technique. *IEEE J. Solid-State Circuits* **SC-19** (December) 828–836.

Shih, C. C. (1985). "Precision Analog-to-Digital and Digital-to-Analog Conversion Using Reference Recirculating Algorithmic Architectures," PhD thesis, UC Berkeley Memorandum no. UCB/ERL M85/60 (July 25).

Subranging A/Ds

Sekino, T., Takeda, M., and Koma, K., (1982). A monolithic 8b two-step parallel ADC without DAC and subtractor circuits. *IEEE ISSCC Dig. Tech. Papers* **XXV** 46–47.

Shimizu, T., Hotta, M., Maio, K., and Ueda, S., (1989). A 10-bit 20-MHz two-step parallel A/D converter with internal S/H. *IEEE J. Solid-State Circuits* **24** (1) 13–20.

Doernberg, J., Gray, P. R., and Hodges, D. A. (1989). A 10-bit 5-Msample/s CMOS two-step flash ADC. *IEEE J. Solid-State Circuits* **24** (2) 241–249.

Kerth, D. A., Sooch, N. S., and Swanson, E. J. (1989). A 12-bit 1-MHz two-step flash ADC. *IEEE J. Solid-State Circuits* **24** (2) 250–255.

Koen, M. (1989). High-performance analog-to-digital converter architectures. Proceedings of the 1989 Bipolar Circuits and Technology Meeting, pp. 35–43.

Tsugaru, K., Sugimoto, Y., Noda, M., Iwai, H., Sasaki, G., and Suwa, Y. A 10-bit 40-MHz ADC using 0.8-μm BI-CMOS technology. Proceedings of the 1989 Bipolar Circuits and Technology Meeting, pp. 48–51.

Ishikawa, M. and Tsukahara, T. (1989). An 8-bit 50-MHz CMOS subranging A/D converter with pipelined wide-band S/H. *IEEE J. Solid-State Circuits* **24** (6) 1485–1491.

Mayes, M. K. and Chin, S. (1989). A multistep A/D converter family with efficient architecture. *IEEE J. Solid-State Circuit* **24** (6) 1492–1497.

Kolluri, M. P. V. (1989). A 12-bit 500-ns subranging ADC. *IEEE J. Solid-State Circuits* **24** (6) 1498–1506.

Hosotani, S., Miki, T., Maeda, A., and Yazawa, N. (1990). An 8-bit 20 ms/s CMOS A/D converter with 50-mW power consumption. *IEEE J. Solid-State Circuits* **25** (1) 167–172.

Chapter 2

CMOS Flash A/Ds

Dingwall, A. (1979). Monolithic expandable 6b 15-MHz CMOS/SOS A/D converter. *IEEE ISSCC Dig. Tech. Papers* **XXII** 126–127.

Dingwall, A. (1979). Monolithic expandable 6-bit 20-MHz CMOS/SOS A/D converter. *IEEE J. Solid-State Circuits* **SC-14** (December) 926–932.

Fujita, Y., Masuda, E., Sakamoto, S., Sakaue, T., and Sato, Y. (1984). A bulk CMOS 20-MSS 7b flash ADC. *IEEE ISSCC Dig. Tech. Papers* **XXVII** 56–57.

Tsukada, T., Nakatani, Y., Imaizumi, E., Toba, Y., and Ueda, S. (1985). CMOS 8b 25-MHz flash ADC. *IEEE ISSCC Dig. Tech. Papers* **XXVIII** 34–35.

Yukawa, A. (1985). A CMOS 8-bit high-speed A/D converter IC. *IEEE J. Solid-State Circuits* **SC-20** (3) 775–779.

Joy, A. K., Killips, R. J., and Saul, P. H. (1986). An inherently monotonic 7-bit CMOS ADC for video applications. *IEEE J. Solid-State Circuits* **SC-21** (3) 436–439.

Kumamoto, T., Nakaya, M., Honda, H., Asai, S., Akasaka, Y., and Horiba, Y. (1986). An 8-bit high-speed CMOS A/D converter. *IEEE J. Solid-State Circuits* **SC-21** (6).

Bipolar Flash A/Ds

Peterson, J. G. (1979). A monolithic, fully parallel, 8b A/D converter. *ISSCC Dig. Tech. Papers* **XXII** 128–129.

Bucklen, W. Monolithic bipolar circuits for video speed data conversion. TRW LSI Publication TP2A-2/81

Muto, A. S., Peetz, B. E., and Rehner Jr., R. C. (1982). Designing a ten-bit, twenty-megasample-per-second analog-to-digital converter system. *Hewlett-Packard Journal* **33** (11) 9–20.

Takemoto, T. and Inoue, M. (1982). A fully parallel 10-bit A/D converter with video speed. *IEEE J. Solid-State Circuits* **SC-17** (6) 1133–1138.

Inoue, M., Sadamatsu, H., Matsuzawa, A., Kanda, A., and Takemoto, T. (1984). A monolithic 8-bit A/D converter with 120 MHz conversion rate. *IEEE J. Solid-State Circuits* **SC-19** (6) 837–841.

Yoshii, Y., Asano, K., Nakamura, M., and Yamada, C. (1984). An 8-bit, 100 ms/s flash A/D. *IEEE J. Solid-State Circuits* **SC-19** (6) 842–846.

Zojer, B., Petschacher, R., and Luschnig, W. A. (1985). A 6-bit/200-MHz full Nyquist A/D converter. *IEEE J. Solid-State Circuits* **SC-20** (3) 780–786.

Hotta, M., Maio, K., Yokozawa, N., Watanabe, T., and Ueda, S. (1986). A 150 mW 8-bit video-frequency A/D converter. *IEEE J. Solid-State Circuits* **SC-21** (2) 318–323.

Goodenough, F. (1986). Flashy 8-bit ADC samples at 25 MHz. *Electronic Design* **34** (May) 65–68.

Peetz, B., Hamilton, B. D., and Kang, J. (1986). An 8-bit 250 megasample per second analog-to-digital converter: operation without a sample and hold. *IEEE J. Solid-State Circuits* **SC-21** (6).

Yoshii, Y., Nakamura, M., Hirasawa, K., Kayanuma, A., and Asano, K. (1987). An 8b 350-MHz flash ADC. *ISSCC Dig. Tech. Papers* **XXX** 96–97.

Akazawa, Y., Iwata, A., Wakimoto, T., Kamato, T., Nakamura, H. and Ikawa, H. (1987). A 400MSPS 8b flash AD conversion LSI. *ISSCC Dig. Tech. Papers* **XXX** 98–99.

Hotta, M., Shimizu, T., Maio, K., Nakazato, K., and Ueda, S., (1987). A 12-mW 6-bit video-frequency A/D converter. *IEEE J. Solid-State Circuits* **SC-22** (6) 939–943.

van de Grift, R. E. J., Rutten, I. W. J. M., and van der Veen, M., (1987). An 8-bit video ADC incorporating folding and interpolation techniques. *IEEE J. Solid-State Circuits* **SC-22** (6) 944–953.

Van De Plassche, R., and Baltus, P. (1988). An 8-bit 100-MHz full-Nyquist analog-to-digital converter. *IEEE J. Solid-State Circuits* **23** (6) 1334–1344.

Wakimoto, T., Akazawa, Y., and Konaka, S., (1988). Si bipolar 2-GHz 6-bit flash A/D conversion LSI. *IEEE J. Solid-State Circuits* **23** (6) 1345–1350.

Lane, C. (1989). A 10-bit 60 MSPS flash ADC. *Proceedings of the 1989 Bipolar Circuits and Technology Meeting.* pp. 44–47.

Mangelsdorf, C. W. (1990). A 400-MHz input flash converter with error correction. *IEEE J. Solid-State Circuits* **25** (1) 184–191.

Chapter 3

Static A/D Performance Parameters

Meyer, P. L. (1970). The method of least squares. *In* "Introductory Probability and Statistical Applications," Section 14.5, pp. 299–302. Addison-Wesley, Reading, MA.

Tewksbury, S. K., Meyer, F. C., Rollenhagen, D. C., Schoenwetter, H. K., and Souders, T. M. (1978). Terminology related to the performance of S/H, A/D, and D/A circuits. *IEEE Trans. Circuits and Systems* **CAS-25** (7) 419–426.

Kulpinski, J. S. *et al.* (1984). Electrical characterization of analog microcircuits. *Rome Air Development Center Technical Report* RADC-TR-83-264 **I** Section VII.

Sabolis, C. (1988). Understanding data converters. *In* "Data Conversion Products 1988-1989," Micro Networks Inc., Worcester, MA.

"16-bit A/D converter performance testing technical note," Bulletin No. 16-100323 Rev. 0, ANALOGIC Corp. Wakefield, MA, (1988).

JEDEC Standard No. 99, Addendum No. 1. "Terms, Definitions, and Letter Symbols for Analog-to-Digital and Digital-to-Analog Converters," Electronic Industries Association, Washington, DC, (July, 1989).

Dynamic A/D Performance Parameters

Gersho, A. (1978). Principles of quantization. *IEEE Trans. Circuits and Systems* **CAS-25** (7) 427–436.

"Understanding flash A/D converter terminology," TRW Electronic Components Group, application note.

"Understanding high-speed A/D converter specifications," Computer Labs, application notes, (1984).

Linnenbrink, T. E. (1984). Effective bits: is that all there is? *IEEE Trans. on Instrumentation and Measurement* **IM-33** (3) 184–187.

Friend, B., Karlak, D., and Sauerwald, M., (1986). Verifying the performance of flash ADCs. *Electronic Engineering Times,* (August 4) pp. T-12 and T-30.

LaBouff M., and Sockolov, S. (1987). Test flash a-d converters to unearth hidden specs. *Electronic Design* **35** (June 25) pp. 119–124.

Swager, A. W. (1989). Flash ADCs push speed and bandwidth limits. *EDN* **34** (May 25) 93–105.

"Dynamic performance testing of A to D converters," Hewlett-Packard Product Note 5180A-2.

Differential Gain and Differential Phase Errors

Felix, M. (1976). Differential phase and gain measurements in digitized video signals. *J. SMPTE* **85** (February) 76–79.

Kester, W. (1978). Characterizing and testing A/D and D/A converters for color video applications. *IEEE Trans. Circuits and Systems* **CAS-25** (7) 539–549.

Chocheles, E. (1984). Increased A/D resolution improves image processing. *Electronics Products* (October 15).

"Comparison of NTSC, PAL and SECAM video levels," Application Note AN-3 Brooktree Corporation, San Diego, CA.

Amorese, P. and Bloomfield, J. (1988). A slew of standards for camera systems. *ESD:The Electronics Systems Design Magazine,* (March) 94–98.

Chapter 4

Chirlian, P. M. (1969). "Basic Network Theory," McGraw-Hill, New York.

Neal, J. and Surber, J. (1984). Flash converters work better with track/holds. *Analog Dialogue* **18** (2) 10–14.

Watson, D. (1985). At video bandwidths, flash a-d converters dictate stringent design. *Electronic Design* **33** (June 6).

Potson, D. (1987). "Track and Hold Amplifiers Improve Flash A/D Accuracy," Application Note TH-06, Comlinear Corporation, Fort Collins, CO.

"Selecting and Using High-Speed Track and Hold Amplifiers," Application Note TH-05 (March, 1987) Comlinear Corporation, Fort Collins, CO.

Givens, M. and Saniie, J. (1987). For fast analog signals build a 50-MHz data acquisition system. *Electronic Design* **35** (July 23) 153–158.

Underwood, B. (1988). High-speed buffers help solve problems in circuit applications. *EDN* **33** (January 21).

Lechner, A., Jessner, H., and Petschacher, R. (1988). Data capture matches flash converter speeds. *Electronic Design* **36** (March 31) 101–106.

Franco, S. (1989). Simple techniques provide compensation for capacitive loads. *EDN* **34** (June 8) 147–149.

Kuzdrall, J. A. (1989). Drive SAR ADCs with low impedance buffer. *Electronic Design* **37** (October 12) 75–82.

Chapter 5

Black Jr., W. C. and Hodges, D. A. (1980). Time interleaved converter arrays. *ISSCC Dig. Tech. Papers* **XXIII** 14–15.

Morgan, D. R. (1982). "Extending the Capabilities of the MC10315/10317 Flash A-D Converters," application note AN-844, Motorola Inc.

Brugemann, H. (1983). Ultrafast feedback A/D conversion made possible by a nonuniform error quantizer. *IEEE J. Solid-State Circuits* **SC-18** (1) 99–105.

Engineering Staff Analog Devices Inc. (1986). "Analog-Digital Conversion Handbook," (3rd edition) Prentice Hall, Englewood Cliffs, NJ.

Poulton, K., Corcoran, J. J., and Hornak, T. (1987). A 1-GHz 6-bit ADC system. *IEEE J. Solid-State Circuits* **SC-22** (6) 962–970.

Demler, M. J. (1988). Understand CMOS flash ADCs to apply them effectively. *EDN* (33) 127–134.

Gee, A. and Young, R. W. (1988). Signal conditioning and analog-to-digital conversion for a 4-MHz, 12-bit waveform recorder. *Hewlett-Packard Journal* **39** (1) 15–22. Hewlett-Packard Co., Palo Alto, CA.

Multistage error correcting A/D converters. *In* "High Speed Design Seminar," pp. I-97 to I-103, Analog Devices Inc., Norwood, MA, (1990).

"Generate non-linear transfer functions with TRW flash A/Ds," TRW application note.

DeLurio, T. "Application Note for two ping-ponged 8-bit flash A/D converters," Signal Processing Technologies application note AN103.

Chapter 6

Gray, G. A. and Zeoli, G. W. (1971). Quantization and saturation noise due to analog-to-digital conversion. *IEEE Trans. on Aerospace and Electronic Systems* (January) pp. 222–223.

Naylor, J. R. (1978). Testing digital/analog and analog/digital converters. *IEEE Trans. Circuits and Systems* **CAS-25** (7) 526–538.

Peetz, B. E. Muto, A. S., and Neil, J. M. (1982). Measuring waveform recorder performance. *Hewlett-Packard Journal* **33** (11) 21–29. Hewlett Packard Co., Palo Alto, CA.

Michaels, S. R. (1984). Watch for superposition errors in data-converter applications. *EDN* **29** (September 20) 255–258.

"Dynamic Performance Testing of A to D Converters," Hewlett Packard Product Note 5180A-2, Hewlett Packard Co., Palo Alto, CA.

Carrier, P. (1983). A microprocessor based method for testing transition noise in analog to digital converters. *Proceedings of 1983 IEEE International Test Conference* Paper 20.6, pp. 610–620.

ANSI/IEEE Std 746-1984, (1984). "IEEE Standard for Performance Measurements of A/D and D/A Converters for PCM Television Video Circuits," The Institute of Electrical and Electronics Engineers, Inc., New York.

Doernberg, J., Lee, H.-S., and Hodges, D. A. (1984). Full-speed testing of A/D converters. *IEEE J. Solid-State Circuits* **SC-19** (6) 820–827.

Mahoney, M. (1985). How to test a flash converter, part 1: fundamental problems. *LTX Technical Topics* **6.1** (November).

Mahoney, M. (1985). How to test a flash converter, part 2: DSP procedures. *LTX Technical Topics* **6.2** (November).

Goddard, T. (1985). "Cross-Plot Generator Allows Quick A/D Converter Evaluation," Precision Monolithics Inc., Application Note 101, (December).

Kuffel, J., McComb, T. R., and Malewski, R. (1987). Comparative evaluation of computer methods for calculating the best-fit sinusoid to the digital record of a high-purity sine wave. *IEEE Trans. on Instrumentation and Measurement* **IM-36** (June) 418–422.

Pinkowitz, D. C. "Histograms Simplify A/D Converter Testing," ILC Data Device Corporation, Application Note AN/L-18.

Harris, S. (1987). Dynamic techniques test high-resolution ADCs on PCs. *Electronic Design* **35** (September 3) 109–112.

Frohring, B. J., Peetz, B. E., Unkricj, M. A., and Bird, S. C. (1988). Waveform recorder design for dynamic performance. *Hewlett-Packard Journal* **39** (1) 39–48. Hewlett-Packard Co., Palo Alto, CA.

Jenq, Y. C., and Crosby, P. (1988). Sine wave parameter estimation algorithm with application to waveform digitizer effective bits measurement. Tektronix, Inc., Beaverton, OR. (Presented at IEEE IMTC/88, San Diego, CA, April 19–22).

Jenq, Y. C. (1988). Measuring harmonic distortion and noise floor of an A/D convertor using spectral averaging. Tektronix, Inc., Beaverton, OR. (Presented at IEEE IMTC/88, San Diego, CA, April 19–22, 1988).

Jenq, Y. C. (1988). Asynchronous dynamic testing of A/D converters. *Handshake* **13** (2) 4–7. Tektronix, Inc., Beaverton, OR.

Knapp, R. (1988). Evaluate your ADC by using the crossplot technique. *EDN* **33** (November 10) 251–262.

Allen, C. (1989). Test ADCs for noise, DNL, and hysteresis. *Electronic Design* **37** (June 22) 93–94.

IEEE Std 1057 (1989). "IEEE Trial-Use Standard for Digitizing Waveform Recorders," The Institute of Electrical and Electronics Engineers, Inc., New York.

Novellino, J. (1990). Product report focus—GPIB control software. *Electronic Design* (February 22) 159–168.

Kester, W. (1990). Flash-ADC testing, part 1. Flash ADCs provide the basis for high-speed conversion. *EDN* **35** (January 4) 101–110.

Kester, W. (1990). Flash-ADC testing, part 2. DSP test techniques keep flash ADCs in check. *EDN* **35** (January 18) 133–142.

Kester, W. (1990). Flash-ADC testing, part 3. Measure flash-ADC performance for trouble-free operation. *EDN* **35** (February 1) 103–114.

Strassberg, D. (1990). Technology update—mathematical software packages. *EDN* **35** (March 15) 53–62.

Schlater, R. (1990). Test fast ADCs with a timing analyzer. *Electronic Design* **38** (April 12) 57–65.

Johnson, M. W. (1990). Test high-speed, high-resolution A-D converters. *Electronic Design* **38** (May 10) 95–100.

"High-Speed Design Seminar," pp. I-51 to I-94, Analog Devices Inc., Norwood, MA, (1990).

FFTs

Harris, F. J. (1978). On the use of windows for harmonic analysis with the discrete Fourier transform. *Proc. of the IEEE* **66** (1) 51–83.

Antoniou, A. (1979). "Digital Filters: Analysis and Design," Chapter 13, McGraw-Hill, New York.

Ramirez, R. W. (1985). "The FFT, Fundamentals and Concepts," Prentice-Hall, Englewood Cliffs, New Jersey.

Pinkowitz, D. (1986). Fast Fourier transform speeds signal-to-noise analysis for A/D converters. *Digital Design* (May) pp. 64–66.

LaBouff, M. and Sockolov, S. (1987). Test flash A-D converters to unearth hidden specs. *Electronic Design* (June 25).

Coleman, B., Meehan, P., Reidy, J., and Weeks, P. (1987). Coherent sampling helps when specifying DSP A/D converters. *EDN* **32** (October 15) 145–152.

Lyons, T. D. (1989). Chirp-Z transform efficiently computes frequency spectra. *EDN* **34** (May 25) 161–170.

Index